Acclaim for **David Abram**

An *Orion* Book Award Finalist

"A truly alchemical book. . . . Those of us who still hope for a revolutionary change in our thinking toward animals, the living land and the climate will welcome this book. Abram is an audacious thinker, a true visionary, and, really, just a damn good nature writer." —*San Francisco Book Review*

"Prose as lush as a moss-draped rain forest and as luminous as a high desert night. . . . Deeply resonant with indigenous ways of knowing, Abram lets us listen in on wordless conversations with ancient boulders, walruses, birds, and roof beams. His profound recognition of intelligences other than our own enables us to enter into reciprocal symbioses that can, in turn, sustain the world. *Becoming Animal* illuminates a way forward in restoring relationship with the earth, led by our vibrant animal bodies to re-inhabit the glittering world." —*Orion Magazine*

"A stunning, compelling journey into embodied, earthly intelligence, *Becoming Animal* is philosophy at its engaging best. Prepare for a wild, profound ride into the essence of the human animal—an essence embedded in communion with the Earth. A must read for anyone concerned about the future of the planet and ourselves." —Kierán Suckling, co-founder and Executive Director, Center for Biological Diversity

"In *Becoming Animal*, David Abram has crafted the rarest of literary gems: a sublime effort combining transcendent prose, lucid insight, and lasting consequence." —*Shambhala Sun*

"Refreshing. [Abram] allows himself to be expansive, sentimental, and more than a little mad. . . ." —*Bookforum*

"If we are to survive—indeed, if we are to stop the dominant culture from killing the planet—it will be in great measure because of brave and brilliant beings like David Abram. This is a beautifully written, deeply moving, and important book."

—Derrick Jensen, author of *A Language Older Than Words* and *Endgame*

"Provocative, boldly recalibrating. . . . A creative and visionary ecologist and philosopher, Abram offers perception-heightening insights into the disastrous consequences of our increasing detachment from the living world as we funnel our attention to the cyber realms. [He] draws on his adventures as an itinerant sleight-of-hand magician to forge an inspirited physics of being. We can't 'restore' nature, Abram writes, without 'restorying' life, hence his prodigious, transfixing, and rectifying 'earthly cosmology.'"

—*Booklist* (starred review)

"*Becoming Animal* brings us home to ourselves as living organs of this wild planet. Its teachings leap off the page and translate immediately into lived experience."

—Joanna Macy, Buddhist scholar and activist

"Without doubt one of America's greatest nature writers, one who ably follows in the footsteps of Muir, Thoreau and Leopold. . . . [A] book of such transformative potential that it needs to be read twice in quick succession to get the full benefit. . . . The language is luminous, the style hypnotic. Abram weaves a spell that brings the world alive."

—*Resurgence* magazine

"This book is like a prehistoric cave. If you have the nerve to enter it and you get used to the dark, you'll discover things about storytelling which are startling, urgent and deeply true. Things each of us once knew, but forgot when we were born into the nineteenth and twentieth centuries. Extraordinary rediscoveries!"

—John Berger, author of *Ways of Seeing* and *Why Look at Animals?*

"Pure enthusiasm drives Abram to explore the yearning of our body for the larger body of Earth. . . . [Abram] brings the magician's sense of mystery and playful surprise. . . . His celebratory embrace of all that surrounds him is refreshing in the extreme."

—*Kirkus Reviews*

"As with many deeply original—and radical—books, this work may startle, even provoke the reader in its electric reversal of conventional thought. . . . This is a portrait of the artist as a young raven, arguing, with all the subtlety of his mind, for the mindedness of the body. An exercise of uncanny imagination."

—Jay Griffiths, author of *Wild*

"This brave and magical book summons wild wonder to remind us who we are."

—Amory B. Lovins, Chief Scientist, Rocky Mountain Institute

"Speculative, learned, and always 'lucid and precise' as the eye of the vulture that confronted him once on a cliff ledge, Abram has one of those rare minds which, like the mind of a musician or a great mathematician, fuses dreaminess with smarts."

—*The Village Voice*

"Abram's prose is lighted from within, happy, solid and clear. . . . [*Becoming Animal*] helps the reader remember his or her place in the larger, luminous world."

—*Los Angeles Times*

"This startling, sparkling book challenges the technological temper of our times by returning us to the animal body in ourselves. Abram shows brilliantly how this body brings us back to Earth in a series of acutely moving descriptions of its polysensory genius. An original work of primary philosophy, it is written with verve, passion, and poetry."

—Edward S. Casey, author of *The Fate of Place: A Philosophical History*

David Abram

Becoming Animal

An Earthly Cosmology

David Abram is an ecologist, anthropologist, and philosopher who lectures and teaches widely around the world. His prior book, *The Spell of the Sensuous: Perception and Language in a More-than-Human World,* helped catalyze the emergence of several new disciplines, including the burgeoning field of ecopsychology. The recipient of a Lannan Literary Award for Nonfiction, David was named by both the *Utne Reader* and the British journal *Resurgence* as one of a hundred visionaries transforming contemporary culture. His writings on the cultural causes and consequences of environmental disarray are published in numerous magazines, scholarly journals, and anthologies. A co-founder of the Alliance for Wild Ethics (AWE), David lives with his family in the foothills of the southern Rockies.

ALSO BY DAVID ABRAM

The Spell of the Sensuous

Becoming Animal

Becoming Animal

AN EARTHLY COSMOLOGY

David Abram

Vintage Books
A Division of Random House, Inc.
New York

FIRST VINTAGE BOOKS EDITION, SEPTEMBER 2011

Portions of this work previously appeared in different form in the following: *Orion*; *Architecture, Ethics, and the Personhood of Place*, edited by Gregory Caicco (Lebanon, NH: University Press of New England, 2007); and *The Child, the Painter, and the Forgotten Life of Things* by David Abram (Deer Isle, ME: Haystack Mountain School of Crafts Monograph Series, 2003).

The Library of Congress has cataloged the Pantheon edition as follows:
Abram, David.
Becoming animal : an earthly cosmology / David Abram.
p. cm.
1. Anthropology—Philosophy. 2. Perception. I. Title.
GN33.A32 2010
301.01—dc22 2009041364

Vintage ISBN: 978-0-375-71369-9

www.vintagebooks.com

Printed in the United States of America
20 19 18 17 16

FOR HANNAH AND LEANDER,

WILD AND LUMINOUS CREATURES

Voice: the breath's tooth.
Thought: the brain's bone.
Birdsong: an extension
of the beak. Speech:
the antler of the mind.

—ROBERT BRINGHURST

CONTENTS

NOTE TO THE READER

The word "earth" appears often in the pages that follow. Today many writers choose to capitalize this term whenever it appears. Such a gesture feels overly facile to me, since it leads us to imagine we are respecting this wild planet, and according it appropriate honor, simply by capitalizing its name. In this work I have generally chosen to keep the term in lower case, in order to remember that the earth is not just the round sphere in its entirety but also—first and foremost—the humble ground beneath our feet, the winds gusting around us, and the local waters flowing through us.

Nonetheless, in a few instances I have found it necessary to capitalize Earth. Such capitalization indicates less a shift in meaning than in tone, occurring only at those moments when a certain emphasis is called for, or when a sort of astonishment breaks through the calm surface of the text. At one point in this volume, even the *spelling* of the name alters, becoming Eairth. If you find yourself perplexed by such variation, unable to discern any clear logic behind it, my apologies: rest assured that it follows a rhyme and reason perfectly evident to the author.

Becoming Animal

INTRODUCTION

Between the Body and the Breathing Earth

Owning up to being an animal, a creature of earth. Tuning our animal senses to the sensible terrain: blending our skin with the rain-rippled surface of rivers, mingling our ears with the thunder and the thrumming of frogs, and our eyes with the molten sky. Feeling the polyrhythmic pulse of this place—this huge windswept body of water and stone. This vexed being in whose flesh we're entangled.

Becoming earth. Becoming animal. Becoming, in this manner, fully human.

This is a book about becoming a two-legged animal, entirely a part of the animate world whose life swells within and unfolds all around us. It seeks a new way of speaking, one that enacts our interbeing with the earth rather than blinding us to it. A language that stirs a new humility in relation to other earthborn beings, whether spiders or obsidian outcrops or spruce limbs bent low by the clumped snow. A style of speech that opens our senses to the sensuous in all its multiform strangeness.

The chapters that follow strive to discern and perhaps to practice a curious kind of thought, a way of careful reflection that no longer tears us out of the world of direct experience in order to represent it, but that binds us ever more deeply into the thick of that world. A way of thinking enacted as much by the body as by the mind, informed by the humid air and the soil and the quality of our breathing, by the intensity of our contact with the other bodies that surround.

Yet words are human artifacts, are they not? Surely to speak, or to think in words, is necessarily to step back from the world's presence into a purely human sphere of reflection? Such, precisely, has been our civilized assumption. But what if meaningful speech is *not* an exclusively human possession? What if the very language we now speak arose first in response to an animate, expressive world—as a stuttering reply not just to others of our species but to an enigmatic cosmos that already *spoke to us* in a myriad of tongues?

What if thought is not born within the human skull, but is a creativity proper to the body as a whole, arising spontaneously from the slippage between an organism and the folding terrain that it wanders? What if the curious curve of thought is engendered by the difficult eros and tension between our flesh and the flesh of the earth?

❦

Is it possible to grow a worthy cosmology by attending closely to our encounters with other creatures, and with the elemental textures and contours of our locale? We are by now so accustomed to the cult of expertise that the very notion of honoring and paying heed to our directly felt experience of things—of insects and wooden floors, of broken-down cars and bird-pecked apples and the scents rising from the soil—seems odd and somewhat misguided as a way to find out what's worth knowing. According to assumptions long held by the civilization in which I've been raised, the deepest truth of things is concealed behind the appearances, in dimensions inaccessible to our senses. A thousand years ago these dimensions were viewed in spiritual terms: the sensuous world was a fallen, derivative reality that could be understood only by reference to heavenly realms hidden beyond the stars. Since the powers residing in such realms were concealed from common perception, they had to be mediated for the general populace by priests, who might intercede with those celestial agencies on our behalf.

In recent centuries, an abundance of discoveries and remarkable inventions have transformed this culture's general conception of things—and yet the basic disparagement of sensuous reality remains. Like an old, collective habit very difficult to kick, the directly sensed world is still explained by reference to realms hidden beyond our immediate experience. Such a realm, for example, is the microscopic domain of axons and dendrites, and neurotransmitters washing across neuronal synapses—a dimension entirely concealed from direct apprehension, yet which presumably precipitates, or gives rise to, every aspect of our experience. Another such dimension is the recondite realm hidden within the nuclei of our cells, wherein reside the intricately folding strands of DNA and RNA that ostensibly code and perhaps even "cause" the behavior of living things. Alternatively, the deepest source and truth of the apparent world is sometimes held to exist in the subatomic realm of quarks, mesons, and gluons (or the still more theoretical world of vibrating ten-dimensional strings); or perhaps in the initial breaking of symmetries in the cosmological "big bang," an event almost inconceivably distant in time and space.

Every one of these arcane dimensions radically transcends the reach of our unaided senses. Since we have no ordinary experience of these realms, the essential truths to be found there must be mediated for us by experts, by those who have access to the high-powered instruments and the inordinately expensive technologies (the electron microscopes, functional MRI scanners, radio telescopes, and supercolliders) that might offer a momentary glimpse into these dimensions. Here, as before, the sensuous world—the creaturely world directly encountered by our animal senses—is commonly assumed to be a secondary, derivative reality understood only by reference to more primary domains that exist elsewhere, behind the scenes.

I do not deny the importance of those other scales or dimensions, nor the value of the various truths that may be found there. I deny only that this shadowed, earthly world of deer tracks and moss is somehow less worthy, less REAL, than those abstract dimensions. It is more palpable to my skin, more substantial to my flaring nostrils, more precious—infinitely more precious—to the heart drumming within my chest.

This directly experienced terrain, rippling with cricket rhythms and scoured by the tides, is the very realm now most ravaged by the spreading consequences of our disregard. Many long-standing and lousy habits have enabled our callous treatment of surrounding nature, empowering us to clear-cut, dam up, mine, develop, poison, or simply destroy so much of what quietly sustains us. Yet few are as deep-rooted and damaging as the habitual tendency to view the sensuous earth as a subordinate space—whether as a sinful plane, riddled with temptation, needing to be transcended and left behind; or a menacing region needing to be beaten and bent to our will; or simply a vaguely disturbing dimension to be avoided, superseded, and explained away.

Corporeal life is indeed difficult. To identify with the sheer physicality of one's flesh may well seem lunatic. The body is an imperfect and breakable entity vulnerable to a thousand and one insults—to scars and the scorn of others, to disease, decay, and death. And the material world that our body inhabits is hardly a gentle place. The shuddering beauty of this biosphere is bristling with thorns: generosity and abundance often seem scant ingredients compared with the prevalence of predation, sudden pain, and racking loss. Carnally embedded in the depths of this cacophonous profusion of forms, we commonly can't even predict just what's lurking behind the near boulder, let alone get enough distance to fathom and figure out all the workings of this world. We simply can't get it under our control. We've lost hearing in one ear; the other rings like a fallen spoon. Our spouse falls in love with someone else, while our young child comes down with a bone-rattling fever that no doctor seems able to diagnose. There are things out and about that can eat us, and ultimately will. Small wonder, then, that we prefer to abstract ourselves whenever we can, imagining ourselves into theoretical spaces less fraught with insecurity, conjuring dimensions more amenable to calculation and control. We slip blissfully into machine-mediated scapes, offering ourselves up to any technology that promises to enhance the humdrum capacities of our given flesh. And sure, now and then we'll engage this earthen world as well, *as long we know that it's not ultimate,* as long as we're convinced that we're not stuck here.

Even among ecologists and environmental activists, there's a tacit sense that we'd better not let our awareness come too close to our creaturely sensations, that we'd best keep our arguments girded with statistics and our thoughts buttressed with abstractions, lest we succumb to an overwhelming grief—a heartache born of our organism's instinctive empathy with the living land and its cascading losses. Lest we be bowled over and broken by our dismay at the relentless devastation of the biosphere.

Thus do we shelter ourselves from the harrowing vulnerability of bodied existence. But by the same gesture we also insulate ourselves from the deepest wellsprings of joy. We cut our lives off from the necessary nourishment of contact and interchange with other shapes of life, from antlered and loop-tailed and amber-eyed beings whose resplendent weirdness loosens our imaginations, from the droning of bees and the gurgling night chorus of the frogs and the morning mist rising like a crowd of ghosts off the weedlot. We seal ourselves off from the erotic warmth of a cello's voice, or from the tilting dance of construction cranes against a downtown sky overbursting with blue. From the errant hummingbird pulsing in our cupped hands as we ferry it back out the door, and the crimson flash as it zooms from our fingers.

For too long we've closed ourselves to the participatory life of our senses, inured ourselves to the felt intelligence of our muscled flesh and its manifold solidarities. We've taken our primary truths from technologies that hold the world at a distance. Such tools can be mighty useful, and beneficial as well, as long as the insights that they yield are carried carefully back to the lived world, and placed in service to the more-than-human matrix of corporeal encounter and experience. But technology can also, and easily, be used as a way to avoid direct encounter, as a shield—etched with lines of code or cryptic jargon—to ward off whatever frightens, as a synthetic heaven or haven in which to hide out from the distressing ambiguity of the real.

Only by welcoming uncertainty from the get-go can we acclimate ourselves to the shattering wonder that enfolds us. This animal body, for all its susceptibility and vertigo, remains the primary instrument of all our knowing, as the capricious earth remains our primary cosmos.

I have no intention with this work to offer a definitive statement, much less a comprehensive one. The complicated and often terrifying problems arising at this moment of the earth's unfolding entail the widest possible range of responses, to which every one of us must lend our specific gifts. I've written this book, a spiraling series of experimental and improvisational forays, in hopes that others will try my findings against their own experience, correcting or contesting my discoveries with their own.

This venture will start slowly, gathering energy as it moves. Simple encounters from my own life—encounters unexpected and serendipitous—will provide a loose, structuring frame for each investigation that follows. The early chapters take up several ordinary, taken-for-granted aspects of the perceived world—shadows, houses, gravity, stones, visual depth—drawing near to each phenomenon in order to notice the way it engages not our intellect but our sensing and sentient body. Later chapters delve into more complex powers—like mind, mood, and language—that variously influence and organize our experience of the perceptual field. The final chapters step directly into the natural magic of perception itself, exploring the willed alteration of our senses and the wild transformation of the sensuous, addressing magic and shapeshifting and the metamorphosis of culture.

Many of our inherited concepts (our ready definitions and explanations) serve to isolate our intelligence from the intimacy of our creaturely encounter with the strangeness of things. In these pages we'll listen close to the things themselves, allowing weather patterns and moose and precipitous cliffs their own otherness. We'll pay attention to their unique manner of showing themselves, attuning ourselves to those facets that have been eclipsed by accepted styles of thinking. Can we find fresh ways to elucidate these earthly phenomena, forms of articulation that free the things from their conceptual straitjackets, enabling them to stretch their limbs and begin to breathe?

The early explorations in this book will soon lead us up against some basic cultural assumptions, forcing us to ruminate a range of reflective questions regarding bodies, materiality, and the language of the sciences, as well as the manner in which our words affect the ongoing life of our animal senses. Such discussions will leave us freer to dance in the later chapters, able to follow our investigations wherever they lead.

Some might claim that this is a book of solitudes. For I've chosen to concentrate upon those moments in a day or a life when one slips provisionally beneath the societal surge of forces, those occasions (often unverbalized and hence overlooked) when one comes more directly into felt relation with the wider, more-than-human community of beings that surrounds and sustains the human hubbub. Awakening to citizenship in this broader commonwealth, however, has real ramifications for how we humans get along with one another. It carries substantial consequences for the way a genuine democracy shapes itself—for the way that our body politic *breathes.*

Why, then, is so little attention paid to the social or political spheres within these pages? Because there's a necessary work of recuperation to be accomplished (or at least opened and gotten well under way) before those spheres can be disclosed afresh, and this book is engaged in that work of recuperation. A replenished participation in the human collective, forging new forms of place-based community and planetary solidarity, along with a commitment to justice and the often exasperating work of politics—these, too, are necessary elements in the process, and they compose a vital part of my own practice. But they are not the primary focus of this volume.[n]

Writing is a curious endeavor, swerving from moments of splendid delirium into others of stunned puzzlement, and from there into stretches of calm, focused craft. The writing down of words is a relatively recent practice for the human animal. We two-leggeds have long been creatures of language, of course, but verbal language lived first in the shaped breath of utterance, it laughed and stuttered on the tongue long before it lay down on the page, and longer still before it arrayed itself in rows across the glowing screen.

While persons brought up within literate culture often speak *about* the natural world, indigenous, oral peoples sometimes speak directly *to* that world, acknowledging certain animals, plants, and even landforms as expressive subjects with whom they might find themselves in conversation. Obviously these other beings do not

[n] The phrase that titles this book, "becoming animal," carries a range of possible meanings. In this work the phrase speaks first and foremost to the matter of becoming more deeply human by acknowledging, affirming, and growing into our animality. Other meanings will gradually make themselves evident to different readers. The phrase is sometimes associated with the late-twentieth-century writings of the French philosopher Gilles Deleuze (1925–1995) and his collaborator, the psychoanalyst Félix Guattari (1930–1992). Like many other philosophers, I have drawn much pleasure from Deleuze's endlessly fecund writings, which are fairly brimming with fresh trajectories for thought to follow. We share several aims, including a wish to undermine an array of unnoticed, other-worldly assumptions that structure a great deal of contemporary thought, and a consequent commitment to a kind of radical immanence—even to *materialism* (or what I might call "matter-realism") in a dramatically reconceived sense of the term. My work also shares with his a keen resistance to whatever unnecessarily impedes the erotic creativity of matter.

Despite the commonality of some aims, however, our strategies are drastically different. (One of my meanders through the backcountry will sometimes cross one of Deleuze's lines of flight at an oblique angle, but our improvised trajectories are rarely, if ever, parallel.) As a phenomenologist, I am far too taken with lived experience—with the felt encounter between our sensate body and the animate earth—to suit his philosophical taste. As a metaphysician, Deleuze is far too given to the production of abstract concepts to suit mine. By choosing for my title a phrase sometimes associated with Deleuze's writing, I nonetheless find myself paying homage to the burgeoning creativity of his work, even as I hope to open the phrase to new meanings and associations.

speak with a human tongue; they do not speak in *words.* They may speak in song, like many birds, or in rhythm, like the crickets and the ocean waves. They may speak a language of movements and gestures, or articulate themselves in shifting shadows. Among many native peoples, such forms of expressive speech are assumed to be as communicative, in their own way, as the more verbal discourse of our species (which after all can also be thought of as a kind of vocal gesticulation, or even as a sort of singing). Language, for traditionally oral peoples, is not a specifically human possession, but is a property of the animate earth, in which we humans participate.

Oral language gusts through us—our sounded phrases borne by the same air that nourishes the cedars and swells the cumulus clouds. Laid out and immobilized on the flat surface, our words tend to forget that they are sustained by this windswept earth; they begin to imagine that their primary task is to provide a *representation* of the world (as though they were outside of, and not really *a part of,* this world). Nonetheless, the power of language remains, first and foremost, a way of singing oneself into contact with others and with the cosmos—a way of bridging the silence between oneself and another person, or a startled black bear, or the crescent moon soaring like a billowed sail above the roof. Whether sounded on the tongue, printed on the page, or shimmering on the screen, language's primary gift is not to *re*-present the world around us, but to call ourselves into the vital *presence* of that world—and into deep and attentive presence with one another.

This ancestral capacity of speech necessarily underlies and supports all the other roles that language has come to have. Whether we wield our words to describe a landscape, to analyze a problem, or to explain how some gadget works, none of these roles would be possible without the primordial power of utterance to make our bodies resonate with one another and with the other rhythms that surround us. The autumn bugling of the elk does this, too, and the echoed honks of geese vee-ing south for the winter. This tonal layer of meaning—the stratum of spontaneous, bodily expression that oral cultures steadily deploy, and that literate culture all too easily forgets—is the very dimension of language that we two-leggeds share in common with other animals. We share it, as well, with the mutter and moan of the wind through the winter branches outside my studio. In the spring the buds on those branches will unfurl new leaves, and by summer the wind will speak with a thousand green tongues as it rushes through those same trees, releasing a chorus of rustles and whispers and loudly swelling rattles very different from the low, plaintive sighs of winter. And all those chattering leaves will feed my thoughts as I sit by the open door, next summer, scribbling and pondering.

These pages, too, are nothing other than talking leaves—their insights stirred by the winds, their vitality reliant on periodic sunlight and on cool, dark water seeping up from within the ground. Step into their shade. Listen close. Something other than the human mind is at play here.

SHADOW

(Depth Ecology I)

Brushing past spruce boughs and ducking the low, brittle limbs of firs, you are walking south along a faint deer trail when it enters a grove of whispering aspens—tall, sun-dappled trunks like elegant giraffe necks leaning this way and that, their heads hidden among the leaves. Your legs carry you upslope through darker, needled limbs and then down as the trail opens abruptly onto the eastern shore of a mountain lake. Keep walking. To your right, the skin of the tarn ripples with a dazzle of sunlight, suffusing the air with reflected rays like an army of swords flourished aloft, their gleams passing through one another, cleaving the depth between you and the rock-ridged slopes rising from the opposite shore. So it's hard to see clearly the grand mountain toward which those ridges ascend, or the trees clustered on those slopes; your eyes feel only a vague intimation of green behind the exultation of light.

The cry of a red-tail hawk echoes off the rock faces—then just the gleaming silence. The silence is not perfect; now and then the faint, papery rattle of a dragonfly sounds near the lake's surface. There's also a suspicion of leaves scattered and rocking on the water, though your eyes can hardly focus on them among all the shifting rays.

To the left, however, your gaze easily marshals itself; here the spruce trunks are etched by the sun streaming through the needles, the bark illumined so crisply that you can feel the texture of the scabbed and shingled surfaces, and so your eyes move among the trunks, translating the patterned light into tactile sensations that ride along your skin.

The afternoon light, like the aspen leaves, is slowly deepening into gold, and soon there's a gilt edge to the grasses and the stones underfoot. Mosquitoes dance above the water, and bees swerve past, drawn to the color of your shirt or the scent of your sweat as you make your way past broad rocks that've been drinking the sun all afternoon. Keep hiking. This air thick with light is an enveloping spell, a trance into which the whole place has now fallen, a viscous state of mind shared by you and the spruces and the bees in this honeyed moment.

And then something shifts: a chill breeze against your face draws you to a different awareness. A swift glance around reveals that the sun is now perched, like a kingfisher, upon the high ridge of the mountain. Stillness everywhere, as though the world itself teeters on the edge; a great transformation is afoot. The breeze stirs . . . and falls silent. Everyone—the dragonflies, the dangling needles, the stones and boulders scattered along the shore—every entity seems to hold its breath. The radiant eye of the kingfisher still watches, but its stare is steadily growing less intense. Everything waits upon the quiet metamorphosis now spreading toward you from across the lake. The sun's gaze grows fainter still; soon it's possible to stare back at it without wincing, as it slowly slips behind the ridge. A final glare, a flaring gleam between two trees silhouetted on that ridge—and then the sun is gone.

Cooler now, to be sure. But not just cooler—there's a new texture to the air, a moistness: water suspended in the medium, unseen, though you can feel its presence as the air washes against your face in waves. And riding those waves, vaguely enticing at first, then too pleasureful to resist: smells! Dark, stygian smells gliding over the rippled glass of the lake to mingle with the aromatic dank of the soil underfoot and the high-pitched scent of the green needles, and a faintly fermented fragrance prying open your nostrils (the recent scat of some creature still steaming in the near woods). There's also the musty decay of a collapsed trunk, and the darkly laughing scent of cool water lapping up against the shore (infused with the chemistry of tadpoles and trout and the tannin of drowned leaves), and a host of other whiffs sometimes merged and sometimes distinct, all sparkling like wine in some part of your brain that had earlier been rocked to sleep by the soporific dazzle of sunbeams, but has now been startled into attentive life by this more full-blooded magic, as though your mammalian intelligence has abruptly dropped anchor and suddenly found itself really here, bodily afoot in these damp woods.

That celestial daze you'd been drifting in has receded to the far periphery of your awareness, the last trace of those ethereal phantasms fading before the robust breath of these deeper colors and obscene odors—chthonic powers whose pungency echoes the sweat and strength of your legs as they push, one after the other, against the ground. In company with the clumped grasses and the frogs, you have crossed a threshold whose influence, although unacknowledged in the current era, remains as potent as ever. And so you have slipped into a different realm, a different mindscape.

You have entered the country of shadow. And a vast and brooding presence that had been hiding, moments earlier, behind the gauze of light is now slowly walking toward you through the clarified air. It is the breathing body of the mountain itself.

One of the marks of our obliviousness, one of the countless signs that our thinking minds have grown estranged from the intelligence of our sensing bodies, is that today a great many people seem to believe that shadows are flat. If I am strolling along a street on a cloudless afternoon and I notice a shapeshifting patch of darkness accompanying me as I walk, splayed out on the road perpendicular to my upright self, its appendages stretching and shrinking with the swinging of my limbs, I instantly identify this horizontal swath as my shadow. As though a shadow was merely this flatness, this kinetic pancake, this creature of two dimensions whom one might peel off the street and drape over the nearest telephone wire.

We identify our shadow, in other words, with that visible shape we see projected on the pavement or the whitewashed wall. Since what we glimpse there is a being without depth, we naturally assume that shadows themselves are basically flat—and if we are asked, by a curious child, about the life of shadows we are apt to reply that their lives exist only in two dimensions.

Suppose, however, that on that same afternoon a bumblebee is making its way from a clutch of clover blossoms on one side of the road to another cluster of blooms in an overgrown weedlot across the street, and that as it does so the bee happens to pass between me and the flat shape that my body casts upon the pavement. The sunlit bee buzzes toward me, streaking like an erratic, drunken comet against the asphalt sky, and then it crosses an unseen boundary in the air: instantly its glow dims, the sun is no longer upon it—it has moved into a precisely bounded zone of darkness that floats between my opaque flesh and that vaguely humanoid silhouette laid out upon the pavement—until a moment later the bee buzzes out the opposite side of that zone and emerges back into the day's radiance.

Although it was zipping along several feet above the street, the bumblebee had passed into and out of my *real* shadow. Its visible trajectory—gleaming, then muted, then gleaming again—shows that my actual shadow is an enigma more substantial than that flat shape on the paved ground. That silhouette is only my shadow's outermost surface. The actual shadow does not reside primarily on the ground; it is a voluminous being of thickness and depth, a mostly unseen presence that dwells in the air *between* my body and that ground. The dusky shape on the asphalt touches me only at my feet, and hence seems largely separate from me, even independent of me—a kind of doppelganger. The apparent gap between myself and that flat swath of darkness is what prompts me, now and then, to accept its invitation to dance, the two of us then strutting and ducking in an improvised pas de deux wherein it's never very clear which one of us is leading and which is following. It is now obvious, however, that that shape slinking along on the pavement is merely the outermost edge of a thick volume of shade, an umbral depth that extends from the pavement right on up to my knees, torso, and head—a shadow touching me not just at my feet, but at every point of my person.

This living shadow is born afresh every dawn, or rather, the shadow is what remains of the night as the night's gloom flees the advance of the rising sun. Visiting a friend's home in the outer suburbs of a large city, I stand at dawn in his yard, patient and still like the trees across the road. Facing away from the sunrise, I see that as the night slips away it leaves behind a slender part of itself, a splinter of dusk reaching from my body to the western horizon—a piece of night that slowly detaches from the mother darkness, gradually gathering itself toward the spot where I stand. This shard of residual night draws in closer to me throughout the morning—its visible edge on the ground slowly congealing, its limbs widening—until I can see by its proportions that it is a clear echo of myself. We mimic one another's movements across the umbral realm that stretches between us. All around me I glimpse other leftover slivers of the night gathering themselves toward other standing bodies—toward trees and telephone poles, toward houses and hydrants and a momentarily pensive squirrel. I continue standing as cars pass and children play. A flock of starlings alight on a telephone wire, whistling and rattling, then move on. Toward noon, I notice that my shadow seems to be seeping into my flesh; by midday it has been almost entirely absorbed through the pores of my skin, and—apart from a smidgen of shade on my northern flank—is nowhere to be seen. The winged insects in the bright air lose nothing of their sunlit radiance as they hum or flutter past. The shadow, my personal night, has been enfolded within me, become indistinguishable from me.

Do we notice this? Do we feel somehow different at high noon, when the darkness has seeped into us? Do we feel the weight of our own shadow, the press of its difficult knowledge against the inside of our torso and skull? Is it the shadow itself that looks out through our eyes at midday? Small wonder that so many traditional peoples give themselves over to siesta, and sleep, for an hour or two at this time, letting their tissues and organs respond to this interior visitation by the night, allowing the many cells or souls within them to be tutored by the darkness that has taken temporary refuge within their flesh. But I will not let myself succumb—not now, for I am waiting to glimpse the night-fragments as they begin to leak out from the trees and from the neighbor's half-repaired motorcycle poised upside down on the ground. I am waiting to feel the quiet release as my shadow slips the confines of my skin and slides its fingers, ever so gently, into the afternoon air.

Ahhh, there. One cheek seems cooler. I look down. On that same side, my body now offers shade to eight or ten blades of grass, and to a beetle swaying near the top of one of those blades, its antennae sweeping the breeze, tasting the microclimate. Within a few moments, four further blades have stepped within the shadow. Then several more.

No longer am I so thoroughly under the scrutiny of the sun. Released from the insistent gaze pouring down from the sky, my left hand flexes and scratches the skin above my knee. The dark reflection on the ground—which I will call my shadowflection—does not register this exchange, which takes place entirely within the bounded volume of my fuller shadow. Slowly, almost imperceptibly, the shadowflection is extending along the ground, as the triangular zone of shelter emanating from my body deepens, lengthens, never violating its Pythagorean proportions, expanding imperceptibly toward the eastern horizon.

I step inside the house to prepare some food—slicing a ripe tomato, interlacing the slices with fresh mozzarella cheese, dribbling olive oil on top to hold the ground pepper that finally falls like darkened snow. The flavors intersect, burst, and blend into one another. Afterward I record the results of my morning's experiment. Several hours later, I walk outside to ponder my shadowflection: a slender giant of a man, honed like a sword, lies prone in the golden light. As the sun sinks behind me, he elongates still further, his shoulders narrowing, his head climbing the wall of a far-off house. And then it has become birdlight—all the wingeds calling and chattering as the giant's contours become blurred, indistinct, granting his penumbral powers to the dusk and the oncoming night.

Has my shadow now dissolved and dissipated? Or is it still present but hidden, swallowed within the wider shade of evening? Or is the evening itself nothing other than a garment woven from all our disparate shadows, from those separate darknesses that walk on their own during the day, yet gather themselves into a common thickness as the sun slides behind the hills? So that our individual shadow, as we have said, is our own private patch of night, torn from the black cloak every morning as we stumble out to greet the day?

For the moment, let's venture simply this: the shadow, this elegant enigma, is always with us. Whether at high noon or at midnight, whether it stands quiet within our skin or envelops us as our milieu, the shadow is an inescapable consequence of our physicality—a disruption of the sun's dominion, a disturbing power that we hold in common with boulders and storm clouds and the corpses of crashed airplanes. There do exist a few members of the bodily community that thrive without the dark companionship of shadows—the various winds, for example, or the pane of glass newly set within the window frame. But for most of us material beings, the shadow is a part of our makeup. Our clearest thoughts are those that know this—those that remember their real parentage in both light and shadow, fire and sleep.

Gloop! The mouth of a trout breaks the near surface of the pond, snatching a fly and sinking back into the green depths. Ripples radiate out from that intersection of worlds, jostling the water striders and a few floating leaves. Stepping close to the edge, you peer through the murk to glimpse any fish that might be visible. A twig snaps under your foot, followed by a faint thump-thump, thump-thump of hooves bounding off into the needled distance. It's a vibration more felt in your torso than heard by your ears. You walk more carefully now, glancing down to avoid the exposed roots and the angled, haphazard rocks with pointillist black and crinkly orange lichens spreading across their mottled facets. Cool air laced with amphibious exhalations slides down the back of your throat; the scent of beaver-gnawed saplings seeps up into your brain. Another trout breaks the surface. The water striders rock back and forth on the undulations.

Earlier the sun was gazing down upon this world, its warmth meeting the inner flame of the grasses, its beams ricocheting off the cliffs and the shifting mirror of the lake. Everything here stood in relation to that white fire as you walked in the dazzle of the gleams, your thoughts lured aloft by the royal request pouring down from the sky. But now that celestial imperative has withdrawn behind the mountain, and each thing's attention seems wholly taken by the life that surrounds it: a spider sussing the breeze that's billowing her web, waiting for a more abrupt tug; the stones hunkering close to the soil; the wind licking the water as it glides over the lake, sliding under roots and through striated furrows in the bark of these trunks, while needles comb the invisible rush. Eyes inhale colors, and your body responds to the pigments of lichen and fungi and cliff with your own ruddy colors, with the hue of your skin and your sweat-stained shirt, the dark undertones in your matted hair. For there is an intimacy here that includes you. It is a commonality that you really notice only when you turn your eyes skyward; high overhead, two ducks flapping southward, their feathers agleam in the golden light, are clearly of another world. They fly in the full gaze of the sun. You realize then that the convivial intimacy of this shadowed world does not extend infinitely upward. It is a bounded realm, a zone that breaks off at some distance overhead, where the sunlight spills over the mountain ridge and illuminates the upper air.

Down here, however, we dwell in a different medium: the coolness and clarity of shadow. A shared influence is apparent among the many beings of this bounded realm. The manner in which colors, sounds, and tastes all step forward into your presence, the way these robust qualities confer and converse with your body—

informing your limbs, for instance, about the waterlogged branches that now dream at the bottom of the lake—all this is a gift afforded by the sheltering shade of the mountain. It is that hulking presence, the many-folded mountain rising in wooded slopes and cliffed ridges from the far shore of the lake, scraped by glaciers and worn down by the winds, laden with ice in its high concavities and layered with stories along every rock slide and precipice—it is *the mountain* that lends its gregarious power to the multiple elements of this place. Whether or not you notice its active influence, it is the mountain that defines the mood of this moment where you stand.

To step into the shadow of this mountain is to step directly under the mountain's influence, letting it untangle your senses as the rhythm of your breath adjusts to *its* breathing, to the style of its weather. To step into its shadow is to become a part, if only for this moment, of the mountain's life. Just as shadows are not flat shapes projected upon the ground (but rather dense and voluminous spaces), neither are they measurable quantities, mere consequences of sunlight and its interruption. Shadows are qualitative attributes of the bodies that secrete them. They are time-dependent realms that change their contours with the hour and the season, momentary life zones where the shadow-casting mountain or boulder or body quietly envelops and gathers a range of other bodies under its sway. The shadow is a bounded space and time wherein the mountain is free to spread out of itself, making itself felt in all its unadorned frankness, drawing a cluster of other entities and elements into a common neighborhood—a zone of alliances and reciprocities enabled by the quiet shelter of the mountain's shade. The power of the shadow is like the action of a benevolent monarch who sets aside the gilt robe that he displays every day to the eyes of the world, donning some drab clothes and slipping out the back door of the palace to wander in the surrounding neighborhood as his own honest, simple self. Despite his humble garb the commoners sense his charisma, fall under the orbit of his royal bearing, and hence a new camaraderie, if only for a few hours, emerges among them. To find oneself in the shadow of a mountain is to abruptly find oneself exposed to the private life of the mountain, to feel its huge and manifold influence on the local world that lies beneath it, to enter the gravitational power of its intelligence, a sagacity no longer dissolved in the dazzling radiance of the sun.

Yet the shadows of late afternoon are very different from those of early morning; the mood, the mode of awareness, the qualities imparted are richly different. Now, for instance, the breeze is dying down, and an ephemeral mist is gathering just above the lake's surface: wisps of vapor hover and drift like spirits. The silence is deeper, fuller; a tiny splash sounds now and then, yet no longer the rattle of dragonfly wings, or the chittering calls of squirrels. The trunks and the cliffs are darkening, the needles losing their distinctness. The shadow of the mountain, this vibrant life zone, is giving way to a power still vaster, darker, more profound. The myriad flows between insects and grass, between soil and stone, hawk and water and cliff, seem to be dissipating—the reciprocities and negotiations between neighbors all gradually subsiding. Sure, there are still encounters and exchanges, yet they no longer compose a tightly woven web. The encounters seem more desultory, more random, the inhabitants of this place by the mountain no longer so intimately engaged with one another. For the mountain's shadow is opening outward, losing its boundaries. As the long-hidden sun withdraws even its residual and scattered light from the sky, the mountain is taken up within the oncoming night, its shadow swallowed now by the darker shadow of the earth.

Night is the name we give to the shadow of the earth. This shadow that eats all other shadows. Slowly the wide planet rolls between our animal bodies and the sun; outlines blur, shapes and colors become uncertain, the proximate world loses its stark reality, while a new depth, gentle and beckoning at first, begins to disclose itself above the trees. A few specks of light appear in the thickening blue, like woodfires in a far-below valley. The bottomland of that valley soon gives way to a deeper canyon, then a chasm, then a fathomless abyss—its immeasurable distances lit by a thousand or ten thousand glimmering stars.

And just as the shadow cast by a mountain, when we step within its bounds, opens us to the brooding intelligence of the mountain, so the mammoth shadow of the earth, as it overtakes us, carries us out of ourselves into Earth's own awareness. It opens us to those beings, daily obscured by the sun-drenched atmosphere, who nevertheless populate the broad expanse that our Earth inhabits: the sibling planets with whom it shares the sun's house; and the countless other homes, some relatively near and most hopelessly far-off, that nevertheless compose, with us, the local neighborhood of the infinite. Or maybe we should speak of those sparkling lights as bodies, solitary yet exuberant lives who communicate via electromagnetic pulses across the inexhaustible deep, bending the fabric of space-time around themselves.

Or have our own bodies now shrunk to the scale of dust grains, and are those stars fresh dewdrops on a cluster of webs being spun by a nest of spiders?

The limitless immensity to which our eyes are exposed makes us dizzy, drunk with the pleasure of having our minds confounded and our ready logics exploded into an array of sparks scattered across the ocean of night. Such ecstasies are not easy for us to stand for very long; our habits of thought call us back to more familiar harbors. Perhaps if we were birds, and space were our medium, then this immensity would not throw us. Or if we were a different mammal—a fox, for instance, our nose tuned to smells that drift in ribbons along the ground—we'd hardly notice that alluring openness overhead, and night would be boon for us. But since we balance on just two legs, our heads are held already in the sky, and so we can't avoid the stunning puzzle posed by the stars. Beyond a certain degree of astonished gawking, our necks begin to hurt, and our legs to buckle; our bodies long to lie down horizontal on the earth. We lend ourselves to gravity, becoming adjuncts of the ground itself. Only by thus renouncing the vertical stance—dropping away our upright individuality and leaning back upon the earth, letting our gaze become the gaze of Earth itself—do we make some sense of the endless depths in which Earth dwells.

For those depths are not *our* habitat; they are Earth's. And so it's only by unfurling our limbs and settling back into the body of Earth that the night sky becomes, for us, a steady comfort and a womb. Such, then, is the spell that the shadow of this planet casts upon our flesh. Sooner or later, we lie down. Eventually, our eyes close. We feed our individual lives back into the wider life of the ground itself. And we sleep.

We sleep, allowing gravity to hold us, allowing Earth—our larger Body—to recalibrate our neurons, composting the keen encounters of our waking hours (the tensions and terrors of our individual days), stirring them back, as dreams, into the sleeping substance of our muscles. We give ourselves over to the influence of the breathing earth. Sleep, we might say, is a habit born in our bodies as the earth comes between our bodies and the sun. Sleep is the shadow of the earth as it falls across our awareness. Yes. To the human animal, sleep is the shadow of the earth as it seeps into our skin and spreads throughout our limbs, dissolving our individual will into the thousand and one selves that compose it—cells, tissues, and organs taking their prime directives now from gravity and the wind—as residual bits of sunlight, caught in the long tangle of nerves, wander the drifting landscape of our earth-borne bodies like deer moving across the forested valleys.

HOUSE

(Materiality I)

My right hand is reaching for a book. Fingers are opening, stretching toward the binding, and suddenly the tome is climbing into my hand. It slips along my fingers and settles back against my wrist and the base of my thumb. My other hand is now prying back the front cover. The book falls open between two pages in the middle. At this point in the story, apparently, things are crawling with snails. The pages are filling up with snails—large ones, with fine spiral shells, tan and pink and spotted brown spirals; they're slowly oozing onto my wrist, some falling onto the pale maple wood of the desk. The whole book is now thick with these slow creatures, and my desk itself is now teeming with them . . .

I awaken, and sit up, the pungent smell of ooze slowly giving way to the light streaming in the windows and the sunstruck sheets tangled around my legs. Grinning dizzily at the fast-fading image of all those translucent, calcite spirals, I swivel my legs out and onto the floor. Yesterday's T-shirt is slung over the corner chair; I pull it over my head, stumble into some stiff new jeans, splash tap water on my face, and wander down to the kitchen, trying to decide between cereal and scrambled eggs. Then I lose track of my thoughts—distracted by a black cricket walking across the tiles—

but my arms decide for me, drawing the oats out of the cabinet and goat milk from the fridge. Soon I'm walking carefully back through the house to my little studio, my forefinger crooked in the handle of a porcelain cup, trying to keep the steeping tea from sloshing onto the floor. In the room that I call my studio, I set the tea on the desk, then turn and shut the door.

This room at the end of the house is dug into the hillside, and so three of its four walls have solid earth behind them, imparting a cavelike feel to the place. Two skylights, and a single window in the outer wall, invite the sunlight into this chamber. On cloudless days, long, rectilinear patches of light creep slowly across the interior walls.

With my partner Grietje and our little daughter, I have been ensconced in this house only a few months—yet already my fingers are well acquainted with these whitewashed walls, smudged here and there where a calendar might've hung, or where someone taped up a drawing and peeled it off when they left. To the right of my desk chair, the wall is mottled and discolored where some moisture seeped through from the packed earth in a previous season. These last months have been void of moisture for far too long; the brief snows of winter melted away all too swiftly, leaving the red earth parched and cracked, the sandstone cliffs thirsty, waiting.

In this earthen room the temperature never rises too high; I run my hands along the cool walls, adding my own smudges to the rest. My chair creaks: it is wanting a change of position, a stretch of its metallic bones. I tilt back on its hind legs until its upper back touches the wall, and it cradles me there. I assembled this chair from metal limbs, struts, and screws, a plywood slat across the seat and a square of padding held in place by the stitched fabric cover I pulled over the torso of the thing. It stretches and shifts as my body shifts within it; it even breathes, subtly, as the day swells with warmth and then relaxes into the night—a companionable friend for my back and my bony ass, and so I try to answer its creaking pleas as best I can, flexing its limbs by leaning from side to side, easing its joints and screws so that it will do the same for me as I sit and ponder within it.

This sitting on chairs is a strange new thing for the primate body—holding our hindquarters away from the ground, our flexible spine suspended in air. Civilized, to be sure. Yet how much more nourishment our spines once drew from their oft-renewed friendship with the ground—planting themselves there, like trees, as we prepared our foods and whittled our implements, squatting on our haunches as we wove patterns into bright cloth and chatted with kin. But now we scorn the ground. Gravity, we think, is a drag upon our aspirations; it pulls us down, holds us back, makes life a weight and a burden.

Yet this gravitational draw that holds us to the ground was once known as Eros—as Desire!—the lovelorn yearning of our body for the larger Body of Earth, and of the earth for us. The old affinity between gravity and desire remains evident, perhaps, when we say that we have *fallen* in love—as though we were off-balance and tumbling through air, as though it was the steady pull of the planet that somehow lay behind the eros we feel toward another person. In this sense, gravity—the mutual attraction between our body and the earth—is the deep source of that more conscious delirium that draws us toward the presence of another person. Like the felt magnetism between two lovers, or between a mother and her child, the powerful attraction between the body and the earth offers sustenance and physical replenishment when it is consummated in contact. Although we've lately come to associate gravity with heaviness, and so to think of it as having a strictly downward vector, nonetheless something *rises up into us* from the solid earth whenever we're in contact with it.

We give ourselves precious little chance to taste this nourishment that springs up into us whenever we touch ground, and so it's hardly surprising that we've forgotten the erotic nature of gravity, and the enlivening pleasure of earthly contact. We spend our days walking not on the earth but on fabricated slabs suspended above office floors and basements; at our desks we ride aloft on our chairs; at night we drift to sleep on the backs of beds neatly raised so we needn't be too close to the ground. If we venture out of doors, it's commonly not to wander on foot; instead, we entrust ourselves to the fiery alchemy of the automobile, whose fevered cylinders and whirling tires loft us speedily to our destination without our needing to touch down on the intervening terrain.

But if our artifacts hold us aloft and aloof from the solid earth, they themselves partake of that solidity and that earthiness; something of the ground passes into the struts and beams of our buildings, spreading into the reclining two-by-fours and the prone slats of maple and pine that lie upon them, spreading up through the clay tiles of the floor, stymied and stifled, somewhat, if the tiles are made of plastic, yet even in that subdued condition leaking up through the legs of our desks and radiating out across the surface to kiss our fingertips where they graze the grain, fortifying our bony elbows when we rest our head upon our open hands. It is a kind of pulse, a dark resonance that sustains and feeds things. We can taste of it in the thickness of our mattress or the fluffed happiness of the pillows as we settle back for the night. The rubber tires of our bicycles may seem impervious to such nourishment, yet even as they roll along the tarmac they're massaging the song forth from its stuck and paved-over stupor, wherein the vitality of stones and the vigor of soil and stem lie stunned and struggling to rise through the black and bituminous carapace.

Certain built structures may invite and enhance the erotic pulse of the ground (like many earthen adobes, and thatch-roofed huts, and even a few unusually eloquent skyscrapers); others may muffle and mute that pulse, but they cannot, I think, immobilize the pulse entirely. Every solid thing, whether a toothpick or a trumpet, a porcelain plate or a helicopter, is fashioned from materials once birthed by the earth. Regardless of how profoundly they have been alchemized in the laboratory, the matter that gleams or sleeps in our creations—the stuff that lends its dull density or its porous whimsy to our tools and our machines, to our chairs and our computer screens—retains some trace of its old ancestry in the wombish earth, some memory of an age when it was not fashioned by an exterior will but bodied forth out of strain and exuberance.

Here in this room, a window-shaped swath of sunlight has by now crept all the way across the floor and, bending itself, is beginning to climb the east wall. Around the window in the opposite wall, a rough-hewn frame of pine stained a deep amber hue looks on, each oval knot in the grain like a wide-open eye—a vortex or whirlpool among the lines that flow in waves and ripples along each cross-sectioned plank until one of these dark, corneal knots forces the fluid grain to bend around it, like space-time around a star.

The undulating grain and the irregularities in the wood beckon gently to my animal flesh, echoing the wavelike lines around my knuckles and the knots in my muscled shoulders, evoking the imperfect and improvisational character of all earthborn beings. It is a character also present in stone, and in the unseen currents rippling the clouds outside that window, and the cracks slowly spreading in the painted plaster of these walls (as gravity quietly invites the house to settle more intimately into the ground). There's an affinity between my body and the sensible presences that surround me, an old solidarity that pays scant heed to our overeducated distinction between animate and inanimate matter. Its steady influence upon my life lies far below my conscious awareness, deeper than the animal sensations that stride along my neurons, beneath even the vegetal sensitivities that rise like sap in my veins. It unfolds in an utterly silent dimension, in that mute layer of *bare existence* that this material body shares with the hunkered mountains and the forests and the severed stump of an old pine, with gushing streams and dry riverbeds and even the small stone—pink schist laced with mica—that catches my eye in one such riverbed, inducing me to clasp it between my fingers. The friendship between my hand and this stone enacts an ancient and irrefutable eros, the kindredness of matter with itself.

Before we moved ourselves to this house among the cottonwoods in the high desert of northern New Mexico, we lived for a time in the mountains far north of here, in a broad valley. The forests were thicker there, the snows more insistent in the winter, the rains more constant in the summer. The house we belonged to then had straw bales inside the thick walls for insulation, a brick floor, and many square-cut, rough-hewn beams of fir running this way and that, rising perpendicular from the middle of the floor and up along the inside walls to support the horizontal crossbeams that sliced through the space above, these in turn supporting several beams that slanted at a 45-degree angle toward the peak of the roof, lofting the high, vaulted ceiling.

That remarkable house appeared quite small from the outside, yet since it was composed primarily of one large room (with assorted nooks and crannies), it had an air of great spaciousness within. A separate loft hid the high ceiling at one end of the house; it was there that we slept, and there that Grietje, assisted by me and a worthy midwife from the Idaho backcountry, gave birth to our little child, Hannah, a week before the winter solstice. On solstice night we bundled Hannah out of doors for the first time, introducing her to the stars and the snow-clad mountains.

Within two weeks, Hannah was accompanying us on our daily skis across the farm fields and into the tree-thick canyons, usually tucked under her mom's fleece jacket with just her tiny blue eyes peeking out from under Grietje's scarf as we glided beneath the conifers and the cottonwoods. Slowly, over the course of several moons, Hannah learned to focus her eyes on shapes closer and farther away; back in the house she loved rolling on the floor with us, and delighted in caressing the walls, the cabinets, and the beams as we carried her near them. The building was transformed by her arrival in its midst, and by the steady curiosity that she lavished upon even its plainest facets—a single brick in the floor, a patch of painted wall. The big room became accustomed to her squeals of pleasure and dismay, and to the effervescent giggles bubbling forth whenever we danced with her in our arms, whirling and bouncing to the rhythms of Baba Maal or Salif Keita pulsing out of our cheap speakers.

One day, sixteen weeks after Hannah's arrival, I drove her and her mother along the back roads and highways to the airport in Idaho Falls, from which they would fly to Belgium to visit Grietje's parents for ten days. After several hugs of goodbye, waving to them as they boarded the plane, I grabbed lunch at a downtown diner and headed back toward our quiet valley, past huge potato farms and the weird mobile irrigation structures that look like dinosaur skeletons decaying in the snowy fields. I was glad for the open prospect of a few days on my own—a chance to catch my breath, and write, without the steady demands and befuddlements of family life. I pulled the car alongside the mailbox at the bottom of our driveway and peered in—a few letters, and several colorful, useless ads from the village supermarket. We'd already protested receiving these ads, at the store and at the post office, but to no avail. I drove up the icy driveway and stepped out onto the snow. I remember well my exhilaration as I stretched open my muscles, the happy anticipation of some solo time. I strode toward the house, trying to walk in the same snowprints that my boots had made that morning, then stepped through the door into our home, grateful for the warmth of its walls and the familiarity of the place.

But as I pried off my boots and slung my pack onto the counter, I noticed that something was amiss. Something about the house was distressed, disturbed. I swung open the door again, wondering if I'd left something outside, but nothing was there. I closed the door, grabbed the new mail, and started toward the couch. And then stopped. The walls, the ceiling, the low tables, and even the windows were all glaring at me. The couch, with its thickly upholstered cushions, was keeping me at arm's length. "The little one's gone," I mumbled quietly.

My words seemed to induce a subtle shift in the comportment of the stairs, and the walls started to sag. The whole interior felt heavy, oppressive—accusation had turned to dejection. "But the baby will be back," I said, to no one in particular, and then louder: "Listen! Hannah will be back. In ten days, she'll be home!"

Straightaway the room lightened up. The furniture relaxed, the ceiling stopped glowering. The structure of the house eased and loosened, the walls and the wooden beams settling back, it seemed, into a resolute and patient waiting. The space no longer felt accusatory; indeed, the space no longer appeared to pay me any attention whatsoever. I sank down into the couch with the day's mail.

It was a fruitful ten days, and when Hannah and her mom returned home, life churned on as before—new discoveries unfolded, new intimacies were forged between Hannah's limbs and the repeating rhythm of the staircase, between the interior of certain cabinets and her questioning gaze.

But it was that weird confrontation with the house upon returning home alone from the airport, that day, that alerted me to the way ostensibly inert things in a home are drawn into new life by an infant's curiosity and pleasure—the way their shapes are received and welcomed by a small child, and enjoyed in their peculiarity and wonder long before those objects are named—before their boundaries become definite, their strict utility and purpose defined. The few friends to whom I related that odd encounter naturally assumed that my experience of the accusatory and dejected house was the result of an imaginative distortion, a projection of my own interior mood upon the plastered walls and the wooden frame of the building. Yet this interpretation seemed far too facile, too easy. The encounter, after all, took me entirely by surprise; my own mood, before I entered the house, was one of confident exhilaration—a halcyon mood that was *interrupted* by the dismay of the walls. And even supposing there was some depth in myself that had already, somehow, been feeling despondent, why was it only the house that took on that mood, and not the automobile as I drove home, or the snowy fields? The notion of "projection" fails to account for what it is about certain objects that *calls forth* our imagination. It implies that the objects we perceive are purely passive phenomena, utterly neutral and inert, and so enables us to overlook the way in which such objects actively affect the space around them—the manner in which material things are also bodies, influencing the other bodies within their ambit, and being influenced in turn.

One's relation to one's house, in other words, is hardly a relation between a pure *subject* and a pure *object*—between an active intelligence, or mind, and a purely passive chunk of matter. Previously I myself might have thought of the relationship in such terms, and perhaps I even reverted to such arrogance as I slipped back into the cluttered life of a first-time father. But a half year after that event, another encounter with the building startled me out of my aloof stance once and for all.

Hannah, almost a year old, had been crawling for many months, and was just then learning to walk, balancing herself by holding on to the edge of a low table as she stepped alongside it, then launching herself from there into the open expanse of the big room, veering from one side to the other. During one of these voyages, the telephone rang. It was the house's owner, the landlord, informing us that he had decided to return to the valley and move back into that house, and so we would have to move out in a matter of weeks.

Crestfallen, Grietje and I slowly began packing up all our clothes, all our pots and our pans, taking over a week to gather and order various stray papers and oodles of unsorted bills, the boxes piling up on the floor. I dusted and swept and disassembled various appliances, then swept again after Hannah upended another container of sundry stuff. All the while I wondered how our departure from the house would disrupt our small daughter's cosmos.

I did *not* ponder what effect our departure might have upon the house itself (I had long since forgotten my experience six months earlier, when I'd come back from the airport). Yet despite all my concerns about Hannah, as the day of our departure drew closer—the day we'd load up the U-Haul with boxes and tables and lamps—it was not my daughter but *I* who became anxious. This house was a good home for us; it didn't seem right to leave. Never mind that we planned to resettle in the Rio Grande valley of northern New Mexico, where we had a clutch of old friends: this house where we'd lived for only a year and a half felt more familiar than those friends. It felt like family. It was the place where Grietje and I had multiplied and *become* a family. The prospect of departing the place vexed me; it seemed all wrong.

A few nights before our projected departure, I awoke in the thick of the dark, startled. Someone was in the house.

I stepped out of our loft and listened from the top of the stairs. Silence. I tiptoed down the steps into a room aglow with pale light:

the full moon was staring through the windows and bouncing its radiance off the walls. The counter surfaces were all luminous and drunk with moonlight. But it was something else: all those wooden beams rising, rough-hewn and sinew-strong, from the floor to the rafters, or else plunging from the rafters to the ground, and those others stretched horizontal across the open space overhead, reaching through the walls to support the eaves outside, and the short, angled struts that bridged the transverse beams to those that bore the vaulted ceiling on their slanting backs—it was all these square-cut beams diving this way and that through the unseen air that shone most vividly in the moonlit room, their cracked and splintered surfaces radiant with a somber fire.

Stalwart individuals, muscular, each with its own unique vector and style—its particular grain undulating like jazz among a cluster of knots, or orbiting a single huge knot where a large limb once grew from the trunk, or else gliding smooth and sleek from one end to the other—each wooden post was suffused with its own singular character. And I realized, somewhat surprised, that I already *knew* these powers, that the fluid lines along each beam, the parallel swerves and ripples in the grain of each post, had become familiar to my senses, somehow, although I'd never paid these sinuous patterns any conscious attention. The beams of this house had been quietly conversing with my creaturely body over the course of the year, coaxing my eyes and my wandering fingers in countless moments of distraction, and I now noticed that I already knew them as individuals—knew them without knowing them, that is, until tonight, when they suddenly broke through the cool callus of my assumptions, forcing me to acknowledge the silent exchange, this language older than words in which my muscled limbs were utterly fluent. That discovery, while standing amid all those wooden beams and the moonshadows they cast upon the floor and the walls, quietly gave me back to myself, rooted my mind back in the loamy soil of my breathing body, and so I could suddenly taste the cool bricks with my feet, and the air currents lambent upon my face, could sense even the wind outside whirling under the eaves and gliding over the roof.

No wonder, then, that I'd become more and more anxious as our departure day approached. It was as though we'd been living for a year and a half in a dense grove of old trees, a cluster of firs, each with its own rhythm and character, from whom our bodies had drawn not just shelter but perhaps even a kind of guidance as we grew into a family. My animal senses had grown accustomed to these beams—to these rectilinear lives whose subtly different dispositions had lent a communal warmth to this structure, instilling the uncanny kinship I'd been feeling with this place. I walked over to each now, in turn, leaning against it or tracing its grain with my fingers, and finally knocking upon it with my knuckles—a clumsy gesture, sure, but an instinctive enough way to make grateful contact with the interior of a piece of wood.

And then I climbed the stairs and went back to sleep.

The next morning I awoke more refreshed than I had in weeks. My limbs were eager to pack the last boxes and load up the U-Haul. We hit the road the next day, although the truck got stuck in the snow before reaching the highway. Two neighbors, on their way to work, stopped and helped dig out the tires. We grabbed some crummy coffee at the town café and rolled out of the valley, arguing about the best route to take, calling and waving to the mountains.

WOOD AND STONE

(Materiality II)

One night, while I was rinsing the dishes, the coyotes started howling nearby—a few of them yip-yipping at first, then a couple soprano howls chiming in, then a whole demented chorus of whooping voices crowding out one another. I turned off the tap and glanced over at Hannah; she was wide-eyed, looking toward the big window from the corner where she'd been ripping the pages out of a new picture book. I took her up in my arms and we stepped out the front door into the darkness, Hannah's legs squeezing my torso in trepidation. The crazed chorus was already subsiding. By the time our eyes adjusted to the dark, the howling had stopped, but we could sense several spirits slipping off between the piñons and the low cedars. The next morning, when we went to investigate, we found several mounds of coyote scat just outside the large kitchen window. The coyotes must've been watching us through the glass.

To a small child, awareness is a ubiquitous quality of the world. We are mistaken when we assume that consciousness is an interior human trait that first unfurls inside the young child, who then learns to attribute that same quality to other persons, and perhaps "projects" that quality onto the surrounding world of things and beings. Rather, the newborn emerges into consciousness as into a new medium. What we later objectify as "awareness" is at first like an anonymous element that defines the very substance of existence, glimmering with strange pleasures and yearnings and pains. Only gradually does a kind of locus begin to appear within this floating field of feeling, an inchoate sense of "here-ness" emerging from the anonymous and omnidimensional plenitude. This crystallizing sense of one's body as a general locus of awareness does not arise on its own, but is accompanied by a dawning sense of the rudimentary *otherness* of the rest of the field of feelings. The earliest experience of selfhood, in other words, co-arises with the earliest experience of otherness. One's own awareness is born of a rift within a more primordial anonymity, as one begins to locate one's sensations in relation to sensations and feelings that are somehow elsewhere, and hence in relation to an awareness that is not one's own, but is rather the rest of the world's.

For a long while, the soft body of our mother, with her pliant breasts and warm belly, her deep eyes and gentle voice, is coextensive with the rest of the world. Slowly, however, this nourishing presence differentiates into a plurality of interpenetrating dimensions, into mothertouch and groundtouch and skyglow and night (into ticklegrass and bath and fathersong), all echoed by a deepening differentiation within oneself, as the body learns to coordinate the actions of its various limbs and to move within the cosmos.

The self begins as an extension of the breathing flesh of the world, and the things around us, in turn, originate as reverberations echoing the pains and pleasures of our body.

So the clustered trees, the bricks in the floor, and the sunlight are not first encountered as inert or insentient presences into which, later, the child projects her own consciousness. Rather, the inwardly felt sentience of the child is a correlate of the outwardly felt wakefulness of the sky and the steadfast support of the ground, and the willfulness of the caressing wind; it is a concomitant of the animate surroundings.

Only much later, as the child is drawn deeply into the whirling vortex of verbal language—that flood of phrases that earlier surrounded her simply as a beckoning play of melodic sounds continuous with the cries of ravens and the rumble of thunder—only then is the contemporary child liable to learn that neither the bird nor the storm are really aware, that the wind is no more willful than the sky is awake, and indeed that human persons alone are the carriers of consciousness in this world.

Such a lesson amounts to a denial of much of the child's felt experience, and commonly precipitates a rupture between her speaking self and the rest of her sensitive and sentient body. Yet the pain of this rupture is quickly forgotten by the speaking self. There are more than enough discoveries and distractions to offset the trauma of this self-estrangement, since accepting and abiding by this odd lesson unlocks the gate to the curious universe that all the grown-ups appear to inhabit.

But the breathing body, this ferociously attentive animal, still remembers.

❧

The foot, as it feels the ground pressing up against it, remembers. The skin of the face remembers, turning to meet the myriad facets, or faces, of the world. The tips of the fingers remember well that each sensible surface is also, in its own way, sensitive. The ears, listening, know that all things speak; they wander and browse in the intimate conversation of the world, and sometimes they prompt the tongue to reply.

Even the eyes know this, that *everything* lives—that the dull or gleaming surfaces they gaze at are also gazing back at them, that the colors they drink or dive into have been longing to swallow them and to taste of their hazel, their bright hints of green.

Our chest, rising and falling, knows that the strange verb "to be" means more simply "to breathe"; it knows that the maples and the birches are breathing, that the beaver pond inhales and exhales in its own way, as do the stones and the mountains and the pipes coursing water through the ground under the city. The lungs know this secret as well as any can know it: that the inward and the outward depths partake of the same mystery, that as the unseen wind swirls *within* us, so it also whirls all around us, bending the grasses and lofting the clouds even as it lights our own sensations. The vocal cords, stirred by that breath, vibrate like spiderwebs or telephone wires in the breeze, and the voice itself, laughing and murmuring, joins its song to the water gurgling under the grate.

Only our words seem to forget, sometimes—or is it the one who speaks and writes them who forgets? The contemporary person sits enveloped in a cloud of winged words fluttering out of his mouth, delighting in their colored patterns and the way they flock and follow one another, becoming convinced that he alone is in blossom—that his skull alone bears the pollen that will fertilize the barren field, that the things stand mute and inert until he chooses to speak of them.

Yet the things have other plans. Bereft of our attentions, their migratory routes severed by the spreading clear-cuts and the dams, their tissues clogged by synthetic toxins leaking into the soils and the waters, they nevertheless carry on. The rising temperatures seem to scorch their surfaces ever more frequently now, yet the things of the world continue to beckon to us from behind the cloud of words, speaking instead with gestures and subtle rhythms, calling out to our animal bodies, tempting our skin with their varied textures and coaxing our muscles with their grace, inviting our thoughts to remember and rejoin the wider community of intelligence.

The child's spontaneous affinity with the objects and entities that surround her is a pleasure to behold, but it remains only an amorphous and tentative solidarity. If it were allowed to unfold throughout the course of childhood, intensifying and complexifying as she herself unfolds through adolescence, this early collusion with things would quietly deepen and mature into a nuanced respect for the manifold life of the world, a steady pleasure in the profusion of bodily forms and the innumerable styles of sentience that compose the earthly cosmos.

For little Hannah, stumbling giddily across the ground, each stone that catches her eye, each bird that swoops or tree that rises up before her is a ready counterpart of herself. If she chooses that tree as a friend for an afternoon, the tree will seem to express various feelings that she feels within herself, to display the sort of awareness with which she is most familiar. Particular plants, specific landforms, and especially other animals seem to incarnate and make visible, for the child, particular impulses that she also senses within herself, empowering the small human to begin to notice and differentiate among various elemental forms of feeling, enabling her to begin to navigate a sea of ambiguous moods, emotions, and impulses whose unruly power could easily overwhelm the child. Hence the centrality of other animals in childhood play across all cultures (including, in my culture, an outrageous array of stuffed rabbits and teddy bears and talking mice), and all the animal protagonists central to the stories and rhymed songs that we first hear and tell in childhood.[n]

If Hannah's fascination and friendship with an apple tree quietly growing in the old orchard is encouraged, and allowed to develop, then the spontaneous reciprocity of her early years, which at first assumes that the tree's experience is akin to her own, will gradually deepen into a discovery that the apple tree's rootedness—its inability to move upon the ground—must grant it a range of sensations very *different* from those that she experiences. Later, an awareness of the way its roots draw water from the soil, and the way its leaves metabolize sunlight, will lead her to acknowledge the tree's still deeper difference from herself—inducing her to stretch her sensorial imagination in order to fathom the odd sensation of drinking sunlight from the air. Perhaps she'll suspect that the pleasure of the sun's warmth on the skin of her shoulders provides a distant entry into the sensations felt by those leaves, or that the lusciousness of sinking naked toes into mud approximates something of what those hidden roots must feel after a summer rain drenches the parched valley. At each step of her inward unfolding she'll discover wider differences between herself and the apple

[n] I am informed here by the lucid research and insights of social ecologist Paul Shepard (1926–1996).

tree, yet she'll measure each difference as a difference in *feeling*, as a strangely different way of experiencing the same sky, the same ground, the same rain that she herself experiences. And so finally in adulthood any tree, for all its strange and unbridgeable otherness, will remain for her an animate and experiencing presence, another being, another shape of sensitivity and radiant life.

Such, at any rate, is the mature fruit of the human child's spontaneous affinity with trees, spiders, stones, and storm clouds, when that seed is allowed to grow and to blossom. Only after such an unimpeded childhood does a grown woman know in her bones that she inhabits a breathing cosmos, that her life is embedded in a wild community of dynamically intertwined and yet weirdly different lives. It is a cosmos no less puzzling, no less fraught with uncertainty and confusion, than the rather complicated world of inert objects and mechanical processes in which so many of us seem to dwell, and yet it is alive—a vibrant play of relationships in which our own lives are participant.

But in a civilization that has long since fallen under the spell of its own signs, the conviviality between the child and the animate earth is soon severed, interrupted by the adult insistence (expressed in countless forms of grown-up speech and behavior) that real sentience, or subjectivity, is the exclusive possession of humankind. This collective insistence could not displace the compelling evidence of the child's direct experience were it not for all the technologies that rapidly come to interpose themselves between the child's developing senses and the earthly sensuous, enclosing her ever more tightly within a purely human realm. The broken bond between the child and the living land will later be certified, and rendered permanent, by her active entrance into an economy that engages the land primarily as a stock of resources to be appropriated for our own, exclusively human, purposes.

Cut off in its earliest stage, dammed up close to its source, our instinctive empathy with the earthly surroundings remains stunted in most contemporary persons. Hence, whenever we moderns hear of traditional peoples for whom all things are potentially alive—of indigenous cultures that assume some degree of spontaneity and sentience in every aspect of the perceivable terrain—such notions seem to us the result of an absurdly wishful and immature style of thinking, at best a kind of childish naïveté. No matter how intriguing it might be to experience the land as animate and alive, we know that such fantasies are illusory, and must ultimately come up against the cold stone of reality. We cannot help but interpret whatever we hear of such participatory beliefs according to our own stunted capacity for empathic engagement with the sensed surroundings—a capability that was stifled in us before it could blossom, and which therefore remains immobilized in us, frozen in its most immature form. Confronted with animistic styles of discourse, most of us moderns can only imagine it as a sort of childlike ignorance, a credulous projection of humanlike feelings onto mountains and rivers, which surely amounts to madness for any adult soul. Rocks alive? Yeah, right!

We fail to realize that such a participatory mode of perception, when developed and honed through the harsh discoveries of adolescence and on into adulthood, will inevitably yield a complexly nuanced and many-layered approach to the world. We fail to recognize that over the course of hundreds of generations, such participation with the enfolding earth will by now have been tuned so thoroughly by both the serendipities and adversities of this world, by its blessings and its poisons, its enlivening allies and its predatory powers, as to be wholly beyond the ken of any merely naïve or sentimental approach to things. Our indigenous ancestors, after all, had to survive and flourish without any of the technologies upon which we moderns have come to depend. It seems unlikely that our ancestral lineages could have survived if the animistic sensibility were purely an illusion, if this experience of the sensible surroundings as sensitive and even sentient were a callow fantasy utterly at odds with the actual character of those surroundings. The long survival of our species suggests that the instinctive expectation of animateness, of an interior spontaneity proper to all things, was a very practical way to encounter our environment—indeed, perhaps the most effective way to align our human organism with the shifting vicissitudes of a difficult, dangerous, and capricious cosmos.

For one cannot enter into a felt rapport with another entity if one assumes that that other is entirely inanimate. *It is difficult, if not impossible, to empathize with an inert object.* One cannot feel or suss out the intention of another creature if one denies that that creature *has* intentions; one cannot anticipate the shifting mood of a winter sky if one denies that the sky has moods, if one begrudges things their their own inherent spontaneity and openness.

When we bring mindful awareness to the simple activity of perception, we may notice that what draws our attention to things—what enables our senses to really engage and participate with them—is precisely the open and uncertain character of those things. An entity that captures my gaze is never revealed to me in its totality; it presents some facet of itself to my eyes while always withholding other aspects from my direct apprehension. I never see a ponderosa pine in its entirety—I see one side of this wide trunk with its fissured bark, while the other side remains concealed. When I walk around to view the other side, that first side is obscured. Nonetheless, I now have a fuller sense of this trunk—although its interior structure remains hidden. The corrugated furrows in its bark seem to recede into that interior; they beckon me closer to touch their various layers and to peer within the deep crevasses. Mmmm—there is a faint scent of vanilla emanating from this particular furrow. As my nose follows the scent closer in, a spiderweb snags on my face and is wrecked as I pull away; now the spider herself is racing up the bark just in front of me, climbing toward the heights. I tilt my head back, tracking the spider with my eyes. I can see only sporadic glimmerings of the upper branches; they're mostly hidden by the spray of needles, as the spreading roots of this pine are concealed beneath the ground. If, seized by an uncontrollable urge to know the whole of this being, I brought over a shovel and began to dig up those roots, I'd be endangering the vitality and beauty of this pine, interrupting the very mystery that draws me back to this huge tree day after day.

Although some aspect of any perceived presence is exposed to my eyes or my flaring nostrils, or to the touch of my fingers, there is always some further dimension that remains hidden. This tension between the apparent and the hidden dimensions of each being beckons steadily to my perceiving body, provoking the exploratory curiosity of my senses. Perception is nothing other than this open-ended relationship—the active allurement of my body by a sapling or a stretch of river, or by the crumbling wall of an old riverside mill—and the consequent reply by my limbs and my listening senses, to which the other responds in turn, disclosing some further aspect of itself to my gaze or my attentive ears. If and when I turn my attention elsewhere, I turn away not from an inert object but from a unique and unfinished way of being, an expressive, enigmatic presence with whom I've been flirting, however briefly.

No matter how long I linger with any being, I cannot exhaust the dynamic enigma of its presence. It is this reticence, the inexhaustible otherness of things, that enables them to hold my gaze, to sustain themselves in my awareness. I can never plumb all the secrets of even a single blade of grass—cannot fathom every aspect of its interior composition, or the totality of the relations that it sustains with the soil and the air. I cannot experience it from every angle all at once. And why not? Because I am not a pure spirit that could penetrate instantaneously every nook and cranny of the thing—because I am not a disembodied mind, for whom the world presents no obstacles and no obscurities—because, that is, I myself am a body, a material being of weight and density like this tree or that stone, and so have my own visible facets and my obscurities (my smooth skin, for example, and my calciferous bones hidden beneath a matrix of muscles), and so can explore the world only from where I stand, can encounter things only from my own thingly position in the midst of them. Because, finally, I am a thing myself, and hence have only a finite access to the things around me.

In fact it is difficult to imagine how a pure, immaterial mind could know even the simplest truth about a chestnut oak or a chunk of marble, since without any body—without any dense or fluid presence, without any eyes or skin or subtle surface—it could never come into contact with that rock. Lacking corporeal sensations, it could neither taste the leaves of the oak nor hear the wind pouring through them, could not feel the tension in the bouncing branches nor indeed "feel" anything at all. *We* can feel the trees and the rocks underfoot, because we are not so unlike them, because we have our own forking limbs and our own mineral composition, because—contrary to our inherited conceptions—we are not pure mind-stuff, but are tangible bodies of thickness and weight, and so have a great deal in common with the palpable things that we encounter.

❦

It follows, of course, that the things of the world have a great deal more in common with *us* than we tend to allow. The affinity is obvious in our relation to other mammals, whose legs, ears, eyes, and overall bilateral symmetry make evident, from the start, our substantial kinship. With trees and shrubs the relation seems far more distant, yet the fact that we acknowledge such beings as *alive* already mitigates that distance. But really now: what of stones—of boulders and mountain cliffs? Clearly, a slab of granite is not alive in any obvious sense, and it is hard to see how anyone could attribute such openness or indeterminacy to it, or why they would want to.

Yet the apparent fixity and inertness of rocks is held in place more by a set of inherited concepts than by any direct, sensory experience of the mineral world. One such concept (as familiar in the modern era as it was in the Middle Ages) is that which opposes the inertness of sheer matter to the vibrance of pure spirit, and holds that sensible matter is vacant and lifeless without the influx of spirit—without this bright fire that descends from a divine realm beyond the physical cosmos to inspire and enliven the drab stuff of the world. This old notion, deeply layered into our Western language, neatly orders the things of the experienced world into a graded hierarchy—"the great chain of being"—wherein those phenomena composed entirely of matter are farthest from the divine, while those that possess greater degrees of spirit are closer to the absolute freedom of God. According to the dispensation of spirit, stones have no agency or experience whatsoever; lichens have only a minimal degree of life; plants have a bit more life, with a rudimentary degree of sensitivity; "lower" animals are more sentient, yet still stuck in their instincts; "higher" animals more truly aware—while humans, alone in this material world, are really intelligent and *awake*.

This way of ordering existence, which depends upon an absolute distinction between matter and spirit, has done much to certify our human dominion over the rest of nature. Although it originates in the ancient Mediterranean and reaches its height in medieval Christianity, this old notion was never really displaced by the scientific revolution. Instead it was translated into a new, up-to-date form by a science still tacitly reliant on the assumption of a limitless human mind (or spirit) investigating a basically determinate natural world (or matter).

Yet as soon as we question the assumed distinction between spirit and matter, then this neatly ordered hierarchy begins to tremble and disintegrate. If we allow that matter is *not* inert, but is rather animate (or self-organizing) from the get-go, then the hierarchy collapses, and we are left with a diversely differentiated field of animate beings, each of which has its gifts relative to the others. And we find ourselves not above, but in the very midst of this living field, our own sentience part and parcel of the sensuous landscape.

Consider a largish rock, about three feet in diameter, its irregular bulk reposing upon the ground. Whether or not such a rock sits on the earth anywhere near where you find yourself reading these words, there *is* such a rock resting just outside the studio where I sit writing them. I am not very sure of its composition. Although it has a similar color to the sandstone rocks that sprawl around it, I know, from close inspection, that it does not have their granular texture, but has a more dense and continuous solidity, riven by jagged edges along its surface. Patches of quartz, dully translucent, are visible here and there, embedded in the pink matrix. I glance at it now through the window, offering its shade to the clumped grasses, implacable in its solidity—a familiar and stable presence amid all my whirling thoughts.

Do you know any such rock? No matter its composition or color, is there not some similar stone with whom you are acquainted, a rock whose stable persistence offers a kind of anchor when worries or dreams threaten to carry you away—a dependable presence that provides, at the very least, a place to lean against and rest your limbs?

After I've been adrift in these words and sentences for a long stretch, I look up and find that rock, once again, poised in the same stance, holding its familiar shape against the trees and the blue sky, although the shadow it casts upon the ground has now shifted. I return to my work. After an hour or so, it calls my gaze once again; the sky behind the rock is now dense with cloud, and so its shadow has been subsumed within the common shade—yet the stone still maintains its earlier stance. I'm suddenly struck by the immense exertion it would take to hold myself in such a stable posture, moment after moment, without flinching. To sustain itself in that position, steadily, so that even now it is still bearing that very shape, and still again, and yet again, hour after hour, and day after day—that must take a lot of effort! Not, of course, that the rock has to deal with the kind of mad restlessness that besets a muscled being like myself. Yet simply to hold itself together in a cosmos that is steadily flying apart, to prevail, year after year, against the suck of entropy, seems already to entail a kind of stubbornness, an obdurate persistence that we miss when we think of "being," or bare existence, as a purely passive state. And so I find myself staring at this rock with a new astonishment—a new appreciation for its compacted energy, the wild activity that it displays by its simple presence.

The age-old bifurcation between spirit and matter prompts us to take bare existence for granted, to assume that a simple material presence—like a stone or a mountain—is utterly passive and inert.

We say that the rock "is" here, that the mountains "are" over there; we use this little verb "to be" countless times every day, and yet we forget that it is *a verb,* that it names an *act*—that simply *to exist* is a very active thing to be doing. By suppressing this activity, by taking "being" entirely for granted, as a purely passive state, we flatten the wild contingency of existence, the uncertainty and risk of the present moment. We lock in the closet our vague puzzlement at finding ourselves here, in this very place, at this very moment of the world's unfolding.

"Why is there something rather than nothing?" is the question that philosophers have used to unsettle our complacency regarding the weirdness of existence, to stir us from our forgetfulness, to reawaken our sense of wonder. Yet it is enough to notice the inherent dynamism of the present moment—to notice that mere "existence" is already an upsurge, and not a blank and passive floating—in order to retrieve the sensuous world from the oblivion to which our concepts too often consign it. A solitary rock or a clear-cut stump is utterly inanimate only as long as "being" itself is taken to be static and inert. Our animal senses, however, know no such passive reality; we've already seen how they perceive things only by interacting with them, by entering into relation with their rhythms of disclosure and concealment, their allurement and their reticence. To my animal body, the rock is first and foremost another body engaged in the world: as I turn my gaze toward it, I encounter not a defined and inanimate chunk of matter but an upturned surface basking in the sun's warmth, or a pink and sharp-edged structure protruding from the ground like the shattered bone of the hillside, or an old and watchful guardian of this land—a resolute and sheltering presence inviting me now to crouch and lean my spine against it.

Each thing organizes the space around it, rebuffing or sidling up against other things; each thing calls, gestures, beckons to other beings or battles them for our attentions; things expose themselves to the sun or retreat among the shadows, shouting with their loud colors or whispering with their seeds; rocks snag lichen spores from the air and shelter spiders under their flanks; clouds converse with the fathomless blue and metamorphose into one another; they spill rain upon the land, which gathers in rivulets and carves out canyons; skyscrapers slice the winds and argue with one another over the tops of townhouses; backhoes and songbirds are coaxed into duets by the percussive rhythm of the subway beneath the street. Things "catch our eye" and sometimes refuse to let go; they "grab our focus" and "capture our attention," and finally release us from their grasp only to dissolve back into the overabundant world. Whether ecstatic or morose, exuberant or exhausted, everything swerves and trembles; anguish, equanimity, and pleasure are not first internal moods but passions granted to us by the capricious terrain.

Such is the terrain sketched and painted by an intense and lonely young Dutchman born to a long line of pastors in 1853. Although his first career was that of a preacher, Vincent's passion could not contain itself in that stance, denying his body while straining toward a beauty beyond the visible; it fell back into the world. And at that moment, as his intellectual faith in a truth behind the sensuous fell away, he found himself abruptly caught up and carried by a faith more implacable than any mere belief: the human body's ancient and inexhaustible faith in the breathing earth, in the whispering leaves, in the meandering river and the night and the goodness of the sun. His senses burst open like sunflowers scattering their seeds; he began to paint the surging world.

In Vincent van Gogh's canvases there is nothing that is not alive. There is no point in the light-filled plenum of sky that does not have its own temporal dynamism, its own rhythm, its pulse. The landscape breathes. And each presence, each clump of soil, each stone and stalk of wheat is in vibrant dialogue with the beings around it.

Emerald-black cypresses dance with the darkening sky, their leaves sending ripples into the night air.

A billiard table holds itself aloof from the other tables and the scattered chairs, each of them on convivial terms with the wooden slats of the floor and with the lamps stuttering their radiance into the room.

Pebbles loll with one another and confer with the thousand blades of grass; rumors rush through those grasses and are passed along, in turn, to the blue mountains and the swaying poplars. A white cloud leans down over those mountains, listening.

A river pours through a ravine, imparting its swirling flux to the roiling cliffs and the flamelike bushes, and to the travelers bearing their backpacks alongside it. Each thing, each being, is in steady intercourse with the entities and elements around it, negotiating its passage and exerting its participation in the ongoing emergence of what is.

Stars spin and stagger the night around them, calling out to one another and to the crescent moon; Vincent's stars are not situated *in* space but actively deploy or secrete the space between them. For there is no prior space, no inert world or backdrop against which the things begin to exist; the cosmos is nothing other than this open and unfolding interchange between the powers that compose it. There is no stone or cloud that is absolved of this passionate activity, no brick or brushstroke that is not participant in the co-creation of the present moment.

Even in his so-called *still-lifes* there is nothing inert or inanimate. The artist's pipe, a plate, two onions, and a box of matches—each being radiates into the world around it, and each is affected (even infected) by the others, as an empty bottle, transparent as an eye, gathers the room into itself.

In another painting a worn pair of shoes, empty of their owner's feet, speak persuasively—by their soft folds and green scrapes, and by the darkness they contain—of that other life that quietly possesses them, and of the undulant land that they walk together. And thus this unobtrusive canvas imparts a sense of simple friendships—the respectful reciprocity between a community and the soil that supports it, the conviviality between two neighbors, even the friendship between two shoes resting upon the tiled floor.

Fiercely intelligent, yet bearing a sensibility far more porous than most, Van Gogh was unable, or unwilling, to abstract his intellect from his body's reality, unwilling to abandon the myriad things, to tame his senses and so to stifle the steady eros between his flesh and the flesh of the earth.

Again and again he slides out of himself, through his eyes, to feel the hunkered silence of the olive groves, and to taste the spreading ecstasy of the leaves as they're slowly lit by the climbing sun. And again and again he is invaded, in turn, by the visible—penetrated by the midday languor of the rolling wheat fields, or by the sullen mood of a neighbor's face. Although he writes often to his brother and a few friends (letters of luminous candor and kindness), it is only in the act of drawing or painting that he is able to give expression to this ongoing intercourse, by offering back to the visible a trace of what the visible steadily pours into his chest.

His paintings, then, are windows through which we look onto an earth no less alive and intelligent than ourselves.

A clutch of Van Gogh's canvases are carefully assembled, now and then, and arranged on the walls of some eminent museum. Crowds of people from diverse backgrounds then journey to that location, month after month, to gaze through these many windows into a cosmos that often frightens them with its intensity, yet where they nevertheless feel oddly, uncannily, at home. Many return to the exhibit over and again. The sheer number of people, young and old, wealthy and poor, waiting to shuffle through the exhibition halls and look upon these ferocious canvases are routinely larger than those drawn to the galleries for any other painter; this has already been the case for many decades.

Yet in his own lifetime Vincent's paintings held little interest for those who viewed them; even at the cheapest prices, his brother Theo was unable to sell them. How, then, do we account for this dramatic metamorphosis in the collective aesthetic of an entire culture? How could it be that a host of vivid canvases painted over the course of a scant eight years (from 1883 until his death in 1890)—canvases that appeared bizarre or unpleasant to most persons who encountered them at that time—became, within sixty years, some of the most beloved paintings of any artist in the long history of the West? What event, what magic, could have so rapidly transformed our seeing, so thoroughly altered the way we looked at the world?

It is a genuine puzzle, and one impossible to answer definitively. Yet we can hazard at least a partial answer. While acknowledging a host of contributory factors (from economic and demographic shifts to the spreading influence of new technologies), it is likely that a decisive factor prompting this drastic change in our collective seeing was our increasing exposure to these very paintings . . .

Vincent's canvases could not have had such a catalytic effect if certain religious assumptions, regarding the distance of humankind from earthly nature, had not begun to break down, and if the more modern separation between "subjects" and "objects" had not started to spring various leaks, allowing a more primordial possibility to make itself felt, however faintly. But in the early decades of the twentieth century, human eyes were already being tutored and transformed by these curious canvases, thickly layered with vehement brushstrokes, as they traveled the world—jolting the vision of numerous artists, sparking new styles and inciting new reactions. Soon reproductions of them began to be disseminated in books, and increasingly as prints and posters hung on innumerable walls. And so, steadily, something opened in our seeing. Minute muscles in our eyes loosened themselves from their static and habitual stance in order to respond to the irresistible lure of these living landscapes, learning to welcome commonplace and taken-for-granted objects—chimneys, weeds, wind-blown branches—now in their own energetic spontaneity and willfulness.

This newly refreshed sense of sight, this way of welcoming the sensuous in all its mysterious and multiform wildness, is the greatest gift of Van Gogh's work. His vision, by virtue of its fierce compassion for whatever was right in front of him, loosens the lock on our own senses. It is a rare person who does not emerge from an exhibit of Vincent's paintings to find that the streetside row of maples flexing their branches and muscling their roots under the city sidewalk are far more earnest and vibrant and, well, *visible* than they were before, or that the previously placid apartment buildings across the avenue are now actively jostling and shoving against one another, vying for sky.

A crow launches from one of the near pines, startling us with the rhythmic thudding of wingfeathers as they paddle the blue air. If you follow its trajectory with your eyes, watching that black shape diminish into the distance, you will notice, way off there, a ribbon of vivid green embedded in the darker green of the needled trees. That ribbon is formed by the bright leaves of cottonwoods growing along the creek that winds through this valley. The current drought means that the streambed will be dry by midsummer, but today—in the middle of spring—there is likely a good flow between the rocks. Let's wander down there: there's a particular site, along that stream, that I've been waiting to show you.

Owch! Damn cactus!!! Take care not to stumble into these low prickly pears—their needles easily penetrate the leather of my old boots. But notice: we're meeting fewer and fewer cacti as we descend to the lowest part of the valley.

Lizards skitter at the periphery of our vision, and a cricket pulse rises from the clumped grasses. The ground is thirsty. Meltwater from the last snows and runoff from the scant rains have carried away layer after layer of topsoil from these slopes, leaving pedestals of earth here and there wherever a bit of grass kept the dirt from washing downhill. In the level spots the dry ground has cracked into a rich, red mosaic of pentagonal and hexagonal tiles. A row of tracks with tight toe-prints lopes across this mosaic—a coyote trotting toward the creek. We follow the tracks through a thicket of piñon pines.

Soon there's a change in the soundscape: the cadence of crickets now interlaced with the fluid trills of tree frogs. And do you catch that new sweetness haunting the air? That's the scent of the cottonwood leaves.

And here at last: the quiet gurgling of water.

We walk upstream along the bank as best we can, dodging

branches, skirting clumps of willows and soggy spots, lulled by the liquid voice rolling over the stones. The dense tangle snags and grabs at our clothes, tripping up our feet. Over and again the frog trills fall silent at our approach—though if we wait, hushed for a few moments, the chorus starts up again louder than ever. The vibration of a hummingbird thrums the air near my ears, then rises above the branches and fades into the near distance. I plunge on, leading clumsily through another thicket of willows. Pricked and scratched, I round a bend in the stream, and am brought to a halt.

Here is the emphatic presence I've been longing to visit since the snows melted. I bow for a moment in greeting, then turn to wait for you. As you step into view, I watch you suddenly stop, and stagger backward a few steps, and hear the sharp, startled exhalation of breath that escapes your lips. I turn to gaze with you.

Rising from the other side of the creek is a huge sandstone cliff, carved into long, lateral striations by centuries of flowing water. The cliff leans far over the stream, eclipsing most of the sky. As your eyes travel up its face, I watch your mouth drop open, and see your knees bend as you drop to a crouch. "Wow," I hear you say.

"Uh-huh." We peer into the sculpted face of this rock, letting its convolutions draw our awareness in curves and swoops across its ruddy expanse. Then we wade across the creek to press our hands against it.

After a while, I break the silence, "It's weird, you know, the way so many people accept the notion that stone is inanimate, that rock doesn't move. I mean, really, this here cliff moves *me* every time that I see it."

You sigh, audibly: "Aw, come off it, Abram, now you're stretching things too far. The so-called movement that you speak of, when you say that 'this rock moves you,' is just a metaphor. It's not a real physical movement in, ya' know, the material world, but only an internal *feeling*, a mental experience that has nothing to do with this actual cliff."

"You must be kidding," I blurt, exasperated. "I mean, how can you say that? I just saw you yourself stagger backward when you first caught sight of this rock face! It was quite obviously a physical movement in the actual, material world, and any bird watching from its perch in those cottonwoods would have to agree with me. You were quite palpably moved. Or do you mean to pretend that you were not?"

"Hmmm . . . Well I guess, in this case, that there was some real, material movement."

"Well, then! D'you still want me to pretend that the rock moves you only mentally, or can we both admit that it is a physical, bodily action effected by the potent presence of this other being? Can we admit that your breathing body was palpably moved by this other body? And hence that you and the rock are not related as a mental 'subject' to a material 'object,' but rather as one kind of dynamism to another kind of dynamism—as two different ways of being animate, two very different ways of being earth?"

You are silent, puzzling. I see you gaze back at the rock face now, questioning it, feeling the looming sweep of its bulk within your torso, listening with your muscles and the quiet composition of your bones for what this old, sculpted presence might wish to add to the conversation. I watch you lie back upon the stony soil, giving yourself to the shelter of the overhanging sandstone, inviting the cool embrace of its shadow. Water drips near your face. The stillness, the quietude of this rock is its very activity, the steady gesture by which it enters and alters your life.

RECIPROCITY

(Knowledge I: Science and Experience)

Consider one of your hands for a moment. Allow your gaze to wander across the open field of the palm, following the creases as they cross one another or converge; let your eyes climb the limber trunks of your fingers. Each of these trunks has a single fold at its base, and two forward folds in its stance—apart from the thumb, whose single hinge makes it a more lumbering being than the others. Allow the thumb to reach across and touch the other fingers, investigating their textures with its tip, as the palm's pliant expanse folds in the middle. Now try exploring that hand with the fingers of your other hand, letting them slide into the steep valleys that separate the first hand's fingers. Taste the pleasure of this contact, the easy resonance between these two exceedingly tactile creatures.

The very organ with which we touch and explore the palpable textures and surfaces of the world, the hand is itself a thoroughly palpable, or touchable, entity. It has its own textured surface like the rippled sand of a beach, or the porous skin of a piece of fruit. This hand that touches things, then, is entirely a part of the tactile field that it explores. It is one of the textured inhabitants of that field, like the velvet moss, the splintered surface of a telephone pole, or the scabrous bark of a white oak near your home.

Wander over to that oak, or to a maple, or a sycamore; reach out your hand to feel the surface of a single, many-pointed leaf between your thumb and fingers. Note the coolness of that leaf against your skin, the veined texture your fingertips discover as they roam across it. But notice, too, another slightly different sensation: that you are also being touched *by* the tree. That the leaf itself is gently exploring your fingers, its pores sampling the chemistry of your skin, feeling the smooth and bulging texture of your thumb even as the thumb moves upon it.

As soon as we acknowledge that our hands are included within the tactile world, we are forced to notice this reciprocity: whenever we touch any entity, we are also ourselves being touched *by* that entity.

And it's not just the tactile sense that exhibits this curious reciprocity. The eyes, for example, these luminous organs with which we hunt the shapes and colors of the visible world, are also *a part of* the visual field onto which they open. Our eyes have their glistening surface, like the gleaming skin of a pond, and they have their colors, like the auburn flank of a horse or a patch of pewter-gray sky. When we stumble outside in the morning, rubbing our eyes free of sleep and gazing toward the wooded hillside across the valley, our eyes cannot help but feel their own visibility and vulnerability; hence our animal body feels itself exposed to that hillside, feels itself *seen* by those forested slopes.

Such reciprocity is the very structure of perception. We experience the sensuous world only by rendering ourselves vulnerable to that world. Sensory perception is this ongoing interweavement: the terrain enters into us only to the extent that we allow ourselves to be taken up *within* that terrain.

On some mornings I step outside before pulling on any socks or sliding my feet into their shoes. The soil presses up against my bare feet and shapes itself to them; the clumped grasses massage and wake up my soles. Sharp pebbles stab the thick skin. Drier, more resistant grasses prick and sometimes break under my weight—

ow!—sending my feet back onto the smoother stones. Pale stones are cool to the toes, dark rocks warmer. My feet receive directives from the ground, turning away from the brown, brittle grasses, seeking the press of those green blades that tickle and play against the callused skin and then spring up again, slowly, after I pass. It feels good to bring my life into felt contact with these other lives, even if only for a moment.

But how does my weight feel to those grasses; how do my steps feel to the terrain itself as I walk upon it? As this question rises, I begin to sense the carelessness with which I'm commonly clomping around, greedily amassing sensations. My legs inadvertently slow their pace as the sensitive presence of the land seems to gather beneath my feet, the ground no longer a passive support but now the surface of a living depth; and so my feet abruptly feel themselves being touched, being *felt,* by the ground. My steps slow down further. Flat rocks and rough rocks, needles cast off by the pines, grit that clings between the toes as they flex against the land: each patch of ground requests a different kind of step, which my legs discover only in the doing. My feet are like ears listening downward, and a dark rhythm rises up into me from this contact—a pulse that slows down and deepens the private beat within my chest.

Try this when the mood strikes you: step out upon the solid earth without the intermediary of a rubber or leather sole—without another creature's tanned hide coming between your flesh and that of the earth. Notice the way your feet pressing against the coarse ground are also met *by* that ground, as your skin is probed by the soil and the pliant, bristling blades. How easy it is to sense that the terrain underfoot is the palpable surface of a living presence, and to allow that depth to feel your steps as you walk upon it! Watch how spontaneously your feet relax their pace in order to respect this odd otherness—in order to reply appropriately to the caress and the steady support of that depth, to avoid insulting the living land with your carelessness. An old, ancestral affinity between the human foot and the solid ground is replenished by the simple act of stepping outside without shoes.

Shoes, of course, are necessary accoutrements of civilization, eloquent in their simplicity; a number of Vincent's canvases pay honor to their humble practicality. But we overuse them, and so forget that our feet—those downturned hands at the end of our hind legs—are sense organs as well as tools for transport. We release our nether appendages from these constraints only at night, when we unlace our artificial hooves and leave them by the bedside, swinging our legs up under the covers. Only then are our feet really free to wander and delve, our toes exulting in the rippling sheets, our soles caressing the cool skin of a lover's thigh. Still, as crucial as it may be for the sustenance of our hearts and for the perpetuation of our species, the gravitational attraction between one naked body and another is but a small reflection of the more ongoing and insistent eros between our body and the earth.

The most intimate contact between the body and the earth unfolds not just at the bottom of our feet, but along the whole porous surface of our skin. For earth is not merely that dense presence underfoot—it is also the transparent air that enfolds us. The space between oneself and a nearby bush is hardly a void. It is thick with swirling currents, adrift with pollens and the silken threads of spiders, a medium instilled with whiffs and subtle pheromones and other messages riding the unseen flows that compose the atmosphere of this breathing world. And we are not just immersed in this invisible medium, we're participant with it, inhaling this mystery through our nostrils and drawing it into our lungs, metamorphosing it there, exchanging certain vital elements for others before we breathe it back into the world. The carbon-rich air we exhale provides ready nourishment for the many plants growing around us. They, in turn, transform the air in the course of alchemizing sunlight into matter, quickening it with the oxygen upon which we animals depend. The atmosphere is a subtle ocean steadily generated and rejuvenated by the diverse entities that dwell within it, a fluid medium of exchange between the plants and the animals and the weathered rocks.

If the air is invisible, it is nonetheless quite *palpable;* we can feel the wind moving against our face and riding the curve of our ears.

We can feel it tugging the hair on our heads, can feel the air part around our wrists as our arms slice their way through its unseen thickness. And just as we can sense the relative speed of its motion, the dryness or dampness of its touch, so we can also sense the breeze sampling *our* qualities as it brushes against us, tasting the intensity of our sweat, the mottled texture of our skin. We can feel it lambent on our shoulders, can feel the breeze's caress on our ankles and its tickling play on our spine as it slides under our billowing shirttails. We can feel the air envelop and settle around us as it takes a mold of our shifting shape. If we assent to its inquiries, responding to its varied expeditions along our surface, then our actions take on a new elegance: the grace that flows from an ongoing and improvised dance with the sensuous medium that enfolds us.

As breathing involves a continual oscillation between exhaling and inhaling, offering ourselves to the world at one moment and drawing the world into ourselves at the next, so sensory perception entails a like reciprocity, exploring the moss with our fingers while feeling the moss touching us back, at one moment gazing the mountains and at the next feeling ourselves seen, or sensed, from that distance . . .

Spring's intensity is finally bursting out of the branches. The scent of apricot blossoms makes my nostrils flare, triggering a faint memory of some giddy childhood transport; I stand still, eyes closed, sipping the sweet air with my nose.

But just then another, darker smell floods in and jumps my undefended neurons, making me wince. I recently began a new compost pile behind a near juniper. Orange rinds, leftovers kept too long in the back of the fridge, lettuce leaves turned brown and oozing, scrapes of this and that all merge into a pungent, slightly acrid rumor that drifts with the breeze. I head uphill and that whispering stench fades into the spice of green needles. A mosquito glides by and then circles erratically back, following a ribbon of scent to where it alights on my sleeve. I brush it free, and it drifts off a short distance, then commences to bounce up and down as if on a string. After a few vertical circuits, it swerves back down and settles on the same forearm, just beyond my rolled-up sleeve. I swivel my arm up to eye level in order to look closely at this creature. The mosquito seems oblivious to my gaze: she's already bracing herself against my arm and piercing my skin with her long proboscis. I restrain an impulse to quash this annoying visitor, aware—for once—that here, too, is reciprocity.

If I'm able to smell the zinging scent of the piñons and the strong fermentation of the compost, it is only because I, too, am a part of the olfactory field—because I have my own musk and effluvium, my own chemical emanations that any mosquito can pick up and follow back to my skin. And since I also take pleasure in sampling the world's tastes—since I can chew and swallow the leaves of coriander and parsley in our haphazard garden, and savor the sour tang of the crusty olive bread cooling on its rack in the kitchen—I know well that I, too, must have my *own* tastes, my delectable flavors, my tartness and bitterness, my dark aftertastes.

As an omnivore, an eater of the flesh of plants and sundry animals, willing to taste almost anything—and thus kindred, in my gustatory and cognitive curiosity, to most other omnivores: to bear and raven and raccoon—I find myself entwined in a great gift economy, wherein each life partakes of other lives and gives of itself in return. No matter how earnestly we humans strive to exempt ourselves from this economy (whether by dissociating from our animal bodies, or trying to eradicate every disease to which we're susceptible, or sealing our remains within lead-lined caskets), we cannot escape our participation in the cycles of exchange. If we ingest the land's nourishment not only through our eyes and ears but also through our hungry mouths—chomping leaves, seeds, and muscles with our teeth, moistening them with our saliva, and swallowing them down into our depths, incorporating the world's flesh into our own—it can only be so because we, too, are edible. Because we, too, are food.

I watch the mosquito's abdomen fill up with my blood. Countless times I have annihilated mosquitoes with a slap against my arm, or else brushed them angrily away. Yet today I just watch, humbled, ashamed to be offering only this tiny sip of my blood in return for the abundant sustenance I draw from the biosphere, yet still glad to confirm my membership in the big web of interdependence, as both eater and eaten.

❧

We can sense the world around us only because we are entirely a part of this world, because—by virtue of our own carnal density and dynamism—we are wholly embedded in the depths of the earthly sensuous. We can feel the tangible textures, sounds, and shapes of the biosphere because we are tangible, resonant, audible shapes in our own right. We are born of these very waters, this very air, this loamy soil, this sunlight. Nourished and sustained by the substance of the breathing earth, we are flesh of its flesh. We are neither pure spirits nor pure minds, but are sensitive and sentient *bodies* able to be seen, heard, tasted, and touched by the beings around us.

There are ways of speaking that honor this consanguinity, and so encourage and enhance the reciprocity between the human animal and the more-than-human land. Yet there are also ways of speaking that implicitly deny this conviviality, styles of speech that stifle the spontaneous participation and exchange between our senses and the sensuous topography.

The familiar discourse of "objects" and objective processes, for example, resolutely holds us aloof from the sustaining earth. It forces us to disengage from our bodily senses and to view this wild-flourishing world as though we were spectators coolly observing it from outside. When we uncritically allude to material nature as a set of inert objects, or even as a clutch of determinate, mechanical processes, we block the perceptual interplay between ourselves and our surroundings. We deny the encounter with brooks, fog-banks, and boulders that hold our attention only by concealing various aspects of themselves from our immediate apprehension, beckoning our bodies into an ongoing interchange. When I talk of the aspen or the granite outcrop as a determinate object, I push into unconsciousness my direct experience of trees and rock ledges, contradicting my carnal awareness of them as ambiguous beings with their own enigmatic ways of influencing the space around them, and of influencing me. I contradict my felt sense of the rising moon as a living power.

And so to speak of enveloping nature in determinate, mechanical terms, or even to write of the environment in a purely functional manner, as "our human life-support system," contravenes and cuts short the conviviality between our animal body and the animate earth. It stifles the spontaneous life of the senses. Our eyes begin to glaze over, our ears become deaf to the speech of tree frogs and the articulations of rain. What we say has such a profound effect upon what we see, and hear, and even *taste* of the world!

As a high school student in the nineteen seventies, I was much impressed by our laid-back physics teacher. Sitting on his desk at the front of the classroom, he told us—wide-eyed—that the apparent solidity and palpability of that desk was merely an illusion, for the desk, in truth, was constituted almost entirely of empty space—vast yawning expanses of emptiness shot through with various infinitesimally small and feverishly whirling particles. I couldn't help but wonder why, then, that piece of furniture still held him aloft, suspended three feet above the classroom floor. I stuck my hand in the air, swooshing it back and forth to catch his attention. "How come, then, you don't fall to the floor?"

"Because my body, too," he announced, "is just a bunch of empty space whizzing with particles!" But this explained nothing whatsoever about why his apparent bulk still seemed to be supported by the wooden tabletop—why, for instance, his ostensibly empty and illusory flesh did not just sink into and merge with the voidness of that table, or why he and the table did not merge with the emptiness of the floor. It explained only why I should no longer trust my senses, and should accept that the abstract dimension of subatomic particles—the esoteric world of electrons and gluons and quarks—was in fact a truer, realer world than the one disclosed by my corporeal senses. For several days, I wandered about in a state of intellectual bedazzlement, aware for the first time that everything I saw with my eyes was only illusion, excited that I, alone among the citizens surging past on the sidewalk, knew the hidden truth. Or at least that I knew someone who did.

Or who said that he did. Gradually, over the course of several weeks, I discovered that a certain kind of provisional trust in the evidence of the senses was entirely necessary to accomplish the simplest tasks—like walking to school, or navigating the overgrown banks of the streambed behind the baseball field. Yet my exploration of that stream no longer held as much fascination for me as it had a few weeks earlier. If the visible world was in truth an illusory effect of realer events unfolding in the subatomic world, then I could no longer really learn anything of significance by gazing at the water striders, or by listening to the syncopated thrumming of the frogs. If I wished to know anything of what was *really* real, I would have to apprentice myself to those scientists who were conversant with that hidden dimension—those whose access to high-powered instruments enabled them to peer behind the perceivable world and glimpse its source.

I began to read feverishly in the arcane world of quantum physics, trying to glean whatever I could. Fascinating at first—yet I soon grew despondent. I was spending all my extra time hunched over articles I'd copied from journals on the squeaking Xerox machine on the back wall of the library, and my muscles were not happy about this. The prescription for my eyeglasses got stronger, while my skin wondered what'd become of the wind that used to explode past my face as I cycled the alleys and narrow woodlands of suburban Long Island. No doubt the wind was still there, but I wasn't—I was nose-deep in my books, trying to decipher a language that had nothing to do, as far as I could tell, with the remembered brook gurgling over the rocks, or the glistening eye of a half-submerged bullfrog. I really missed those frogs. *So what* if I couldn't yet gain access to the subatomic world with its ultimate truths? It soon dawned upon me that it was biology I should be studying.

The following year, I gave myself to biology with a vengeance.

This was the science for someone like me—a shy and dreamy kid most at ease when balanced on a high branch in the backyard, listening to the whistling and liquid speech of the local birds, trying to fathom what they were talking about . . . It was good to be back in the same world as the jays and squirrels. It puzzled me at first that biology was taught entirely indoors, in a laboratory-style classroom—but I secured a desk close to the window, and when during lectures my attention drifted out beyond the glass, I felt less guilty about it than I did in my other courses. This being biology, I felt more justified gazing at the underbelly of a beetle climbing the other side of the pane, or a couple of barn swallows swerving and dipping behind the teacher's reflection.

But then, a month and a half into term, Mr. Warasila announced that we'd be dissecting frogs the next week.

"Um, where're these frogs gonna come from?" I demanded of him after class. He told me they were being shipped over for the express purpose of being dissected.

"Are you sure they're not from around here?" I asked.

"Positive."

All right. I ran across the field as soon as school was out, ducked under a few trees, and sat down by the dwindling stream. I wanted to consult the local frogs about this new situation. I sat there, I remember, for a long time, but none of the big-eyed critters turned up. Until much later, at dusk, I heard first one frog, and then another, but there wasn't enough light to locate them with my eyes. So I just sat there and started talking to them, asking if they thought it was okay to dissect some other frogs in class. Whenever I started talking they quieted down, and then after I stopped there'd be a long pause, as though they were maybe thinking about what to say, and then they'd start croaking again. I knew full well that they were not replying to what I'd said, yet it had the rhythm of a conversation, and so I listened close to their resonant calls, and then cut in and tried to justify what I'd be doing in the laboratory next week, and they'd fall silent, once again, at the sound of my voice, and so we went back and forth for a while. I guess I came away feeling that, well, maybe it was okay to dissect some of their distant relatives, since it was necessary in order to learn the Truth about the world.

The following semester, however, our biology class began to focus upon the molecular basis of inheritance. It was then and there I learned a key part of that Truth: that not only the anatomical structure but even the *behavior* of all these animals was "programmed in their genes."

Programmed! This piece of knowledge, as it sank in, reverberated through my own organism, transforming my experience of the organic world. After the last of the snows had melted away that spring, far fewer songbirds came to alight in the backyard branches. Earlier springs had brought an abundance of bird-speech bubbling back and forth between the trees and the hedges; now there seemed little of such chatter. Until one weekend, while sitting on the stoop gazing around, I abruptly realized that there were plenty of birds out and about, but I was no longer *noticing* them, was having a hard time hearing, or focusing in on, their vocalizations. The songs of the thrushes, the loopy cries of cardinals and blackbirds and other wingeds that previously held me rapt with fascination, had begun to fade from my awareness . . . It was as though my ears were becoming impervious to all these non-human voices. My auditory attention was no longer engaged by these various sounds, my listening no longer compelled by whatever was being said by those birds—precisely because, according to my new and improved understanding, they were no longer really *saying* anything at all. Their utterances were just automatic sounds—"programmed," as it were. Sure, some of these repeated sounds were pleasant enough to attend to for a moment (like background music playing on a radio). But because I had defined away all the creativity and meaning in those voices, because there was no longer anything actively *being said* by those birds, my ears found little purchase in those sounds, no enigmatic openness, or otherness, to compel my listening. And so those avian calls readily faded into the background of a life increasingly focused on purely human concerns.

No longer enveloped by a many-voiced world at least as alive and mysterious as my own sentient self, my animal senses began to shut down. My eyes were no longer dazzled by the turquoise beetles climbing the weeds by the train tracks, or by the high-stepping poise of the heron at the local marsh. As I reflect upon it now, it seems that my skin itself became less porous, less permeable to the abundant life that surrounds, as my conscious self steadily withdrew its participation from sensuous nature and began to live, more and more, in a clutch of heady abstractions.

This retreat from directly experienced reality accelerated in college. The human senses, as we were taught in our college science classes, were deceptive; they were not to be trusted. For the genuine reality of things was inaccessible to our unaided senses. The real truth was always hidden behind the scenes—whether tucked inside our skulls (in that dimension of firing neurons and neurotransmitters that ostensibly determines all our experience), or hidden within the nucleus of our cells (in the molecular realm of the nucleotide sequences, or genes, that presumably cause all our curious behaviors), whether in the supersmall world of electrons, neutrons, and quarks, or in that dimension—inconceivably distant in space and time—wherein the cosmos emerged from the primordial "big bang." The world accessible to our senses, the visible world of hillsides and rain and flocking birds, came to seem a secondary dimension, a largely illusory field of appearances waiting to be penetrated, and dissipated, by the human mind. The animate nature that our senses revealed was no longer fundamental; hence few people seemed very upset about the rapid destruction of forests and wetlands, or the accelerating extinctions of diverse creatures. Observable nature was a derivative reality—a realm useful for its extractable resources, or as a dump site for the toxic byproducts of human progress, but not worth worrying about, really, since we could replicate whatever we wish of this domain with our virtual technologies, and would soon engineer the rest of nature, using gene splicing and nanotechnology, to suit our desires.

Such was the heady atmosphere in which my fellow pre-medical students and I were immersed. Few of us understood, back then, that all these technological dreams—like all those abstract dimensions hidden behind, beyond, or underneath the apparent surroundings—were still, secretly, rooted in the immediate world of direct experience, and hence were dependent upon our everyday encounters with the local earth. We did not suspect that our instinctive awareness of the winds, the waters, and the soil underfoot provided the necessary ground for all those abstractions, the sole guarantee of their coherence. It is only now, as we find both our lives and our high-tech laboratories threatened by severe fluctuations in the weather, as we watch coastlines disappear and foodwebs collapse and realize that our own children will not be exempt from the violence that our onrushing "progress" has inflicted upon the earth, only now do we notice that all our technological utopias and dreams of machine-mediated immortality may fire our minds but they cannot feed our bodies. Indeed, most of this era's transcendent technological visions remain motivated by a fright of the body and its myriad susceptibilities, by a fear of our carnal embedment in a world ultimately beyond our control—by our terror of the very wildness that nourishes and sustains us. To recognize this nourishment, to awaken to the steady gift of this wild sustenance, entails that we offer ourselves in return. It entails that we accept the difficult mystery of our own carnal mortality, allowing that we are bodily creatures that must die in order for others to flourish. But it is this that we cannot bear. We are too frightened of shadows. We cannot abide our vulnerability, our utter dependence upon a world that can eat us. Vast in its analytic and inventive power, modern humanity is crippled by a fear of its own animality, and of the animate earth that sustains us.

So those ways of speaking that refer to nature only as a set of determinate objects, and random or mechanical processes, steadily isolate us from our senses and from the sensuous surroundings. They deepen the distance and opposition between ourselves and the grasses, between our conscious reflections and the gusting winds that bend and sometimes break those grasses.

Yet there are *other* styles of discourse that stir the senses from their slumber, other ways of speaking that can resuscitate the forgotten solidarity between the human animal and the animate earth. Words and phrases that respect the ultimate unknowableness of things—not only of other people, but of turtles, and juniper pollen, and the shifting progress of a rainstorm; ways of speaking that accord a certain enigmatic otherness, and uncertainty, even to shed antlers, and rocking chairs, and switched-on lightbulbs that startle us with their sputtering—to the moon, of course, but also to dwindling rivers, and well-made bicycles, and even to words themselves—such are the kinds of speaking that wake up our skin, calling us back inside the body's world.

There are innumerable distinctions to be drawn between the palpable phenomena of this world, yet each particular presence partakes of a common mystery: the unfathomable upsurge of existence itself. Each thing expresses this mystery in its own manner and style, yet each is equivalently outrageous, a clump of dirt no less than a roaring, marauding brown bear—each enacting its own tenuous and improvised way in the world, each gifting its own rhythms to the riot of life that surrounds it. Every gust of wind, every note ringing forth from the bell tower, each staccato step of a water strider along the stream's surface, has its own subtle influence upon the beings around it. Simply to exist, or continue existing, is already active—already a doing—and hence no phenomenon is utterly passive, without efficacy or influence. To allow this influence into our speaking, to affirm the unique dynamism of the various entities we meet in the course of a day, to acknowledge that whatever snags our attention has its own agency, or life: such a way of speaking inevitably begins to open our stifled senses.

Why is this so? Why should such an animistic style of speaking—one that assumes some modicum of creativity in even the most obstinate of phenomena, and which therefore speaks of things not merely as objects but as animate *subjects,* as living powers in their own right—why should such a way of talking renew and rejuvenate our bodily senses?

First, because it opens the possibility of interaction and exchange, allowing reciprocity to begin to circulate between our bodies and the breathing earth. If we speak of things as inert or inanimate objects, we deny their ability to actively engage and interact with us—we foreclose their capacity to reciprocate our attentions, to draw us into silent dialogue, to inform and instruct us. As soon as we allow things their own enigmatic openness and otherness, then our sensing bodies find themselves accosted, affronted, massaged, and ensnared by a host of charismatic powers vying with one another for our attentions. We're suddenly surrounded by a crowd of alluring beings, some of them shy and some of them shameless, each of which provokes the imagination of our eyes or the curiosity of our ears, coaxing our senses into a new conviviality with the local earth.

Second, such a language makes evident the consanguinity between ourselves and the enfolding terrain, invoking an explicit continuity between our lives and the vitality of the land itself. It implies that the creativity we find in ourselves has its correlation in the surrounding cosmos, and, too, that the relative stubbornness and solidity we associate with the things around us have their correlate in the weight and inertia of our own lives, in the density of our flesh and the intransigence of our habits. By implying that each mountain, each cloud, each wolf or oak or hive of bees, is a distant variant of our own texture and pulse, and, conversely, that our own sentient organism is itself a variant of these things—an intensification or fluctuation within the sensitive flesh of the world—such a way of speaking situates the human intellect back *within* the sensuous cosmos. It subverts the long isolation of the thinking self from the perceptual world that it ponders, suggesting that we and the sensorial surroundings are woven of the same fabric, indeed that we are palpably entwined with all that we see, and hear, and touch—entirely a part of the living biosphere.

Third, by describing the myriad things as unfolding, animate *beings,* we bring our language back into alignment with the ambiguous and provisional nature of sensory experience itself—with the fact that we never perceive any entity in its entirety, but only encounter partial aspects according to the angle or mood of our approach. We may *conceptualize* the eroding cliffs and the bracket fungi as determinate phenomena, as fixed and finished presences resting complete in themselves, but we can never really *perceive* them as such. We cannot experience any entity in its totality, because we are not pure, disembodied minds, but are palpable bodies with our own opacities and limits. We are in *and of* the world, materially embedded in the same rain-drenched field that the rocks and the ravens inhabit, and so can come to knowledge only laterally, by crossing paths with other entities and sometimes lingering, responding to a thing's sparkle or its calloused coolness, slowly becoming acquainted with its characteristic tenor and style, the unique manner in which it resists our assumptions. *All* our knowledge, in this sense, is carnal knowledge, born of the encounter between our flesh and the cacophonous landscape we inhabit. We know the things of our world not, in truth, as determinate objects, but as acquaintances and strangers, as trusted familiars and as troublesome neighbors, as allies and misfits and moody, dangerous comrades. Some, like the night sky, command my stunned silence and awe. Others, like my house, I tend to take for granted, although it steadily shelters me, and holds my toothbrushes. Still others, like the pages of a manuscript I'm slowly writing, are both alluring and inscrutable; I know not whether their apparent aloofness is because they disdain me or because they are shy, whether I should approach them or keep my distance. All of these beings are participant, with me, in the ongoing emergence of the real. Not one of them, in my direct experience, is utterly inert or inanimate.

Or is it? What of the plastic spatula in the sink, or the tax return we have yet to fill out this year, or the computer dozing, unplugged, by the wall? I mean, really, let's face it: that computer is not, in actuality, really *sleeping*—it's just turned off, and hence not computing—not really functioning at all, at the moment, except as a fancy paperweight, with all its complex and compacted circuitry immobilized and, well, inert. It is just *there,* an ingeniously fashioned mass of plastic and silicon that could never have been built were it not for the cool detachment of the analytic mind with its propensity to step back from the world, to hunt for underlying mechanisms, to objectify and tinker with things. An understanding of the world in analytic and objectlike terms is part and parcel of our reality; it is by now built into many of the artifacts we use, and many of the institutions we rely upon. It seems unlikely, therefore, that the detached, objectifying stance toward the things and processes of this planet will dissipate anytime soon.

And for heaven's sake, why *should* it dissipate? Think of the astonishing insights and capabilities that this stance has brought us: the chemical table of the elements, the power of rapid transport and even flight, instantaneous communication links to almost anywhere on the planet! Not only technologies that might deaden the mind or deplete the earth, but many sustaining ones as well: the conquest of debilitating diseases, bioremediation of toxins, sophisticated solar and wind-driven engines, my eighteen-speed bicycle . . . Isn't there something worthy and right, then, about our objective understanding of nature, and about the manifold insights that have been garnered from that perspective, the many useful facts and explanations harvested from the careful practice of our sciences?

Certainly there is! Nothing I write in these pages is meant to disparage the elegant disclosures revealed by our sciences, or even the detached states of mind necessary to derive those insights. I mean only to point out that the detached stance proper to science is itself dependent upon a more visceral reciprocity between the human organism and its world. It is this primordial and ongoing participation that has gradually shaped the glistening eyes with which we now view the lights and shadows of this sphere, as it has shaped our listening ears, able to detect a wide range of sounds vital to human life while oblivious to many others. This ageless intercourse between the body and the earth—this *coevolution*—has shaped the organs and tissues of every earthly organism, deepening the color of our feathers and the power of our claws, intensifying the bitterness of our leaves, outfitting our seeds with cotton sails so they'll blow far on the summer winds.

The practice of science may enable us humans to provisionally step aside, now and then, from our immediate involvement in the coevolutionary exchange, yet our participation with the rest of the animate landscape has formed the very eyes that now peer through our microscopes, the very intelligence that now interprets the data.

The intrepid investigations of our sciences have yielded a range of insights that could, rightly construed, be of benefit not just to humankind but to the whole of the biosphere. In order to attain these insights, however, our scientists often had to feign a God's-eye view of the world, assuming a detached and disaffected perspective—pretending, for the duration of their investigation, that they were not themselves entangled in the very nature that they were studying. During the course of each experiment, the scientist necessarily accepts this useful fiction, pretending that his or her observing intellect can somehow break itself free from its coevolved, carnal embedment in the thick of earthly life. She assumes that her intellect can somehow liberate itself from the physical biases and unconscious proclivities of her particular species, in order to observe whatever unfolds as a wholly uninvolved spectator—as an utterly objective rationality, a pure mind.

As a temporary and provisional stance, the assumption of a sheer objectivity has been immensely fruitful; it has resulted in a remarkable range of useful evidences. But while the individual practitioners of science must ultimately revert to their everyday lives as participants in the terrestrial world that they study, the lingo in which their findings are couched remains in the detached and disembodied mode. Their experimentally derived insights, that is, are preserved in the highly abstract discourse necessary to communicate those insights to other researchers feigning the same detached, or severed, state of mind. There is rarely any clear *translation* back from the pose of a pure, bodiless rationality to a language appropriate to our lived engagement and rapport with the life around us.

Any such translation would need to render those insights in a manner accessible to felt, bodily experience. Still more important, it would need to correct for the distortions induced in the interpretation of the raw data by the provisional pretense of a disincarnate perspective, of a view from nowhere.

Since no such translation is available (since each scientific investigation builds on previous investigations without translating the results of those prior studies into the language of direct experience), a host of abstract, provisional worlds is gradually built up by our diverse sciences, a multiplicity of arcane dimensions hidden *beyond, behind,* or *beneath* the experienced world. Such are the quantum dimensions explored by high-energy physics, or the realm of gene sequences pinpointed along the complexly folded strands of DNA tucked within the nuclei of our cells, as well as the tangled forests of axons and dendrites located behind our brows, and even the supervast precincts of black holes and galactic clusters. These recondite realms are largely hidden from our direct experience, and so we must rely on various experts to mediate and interpret the deep matters that are found there. Increasingly, those transcendent dimensions are purported to hold the fundamental truths that ostensibly cause, "code for," or otherwise *explain* the ambiguous, uncertain world of our direct experience. The quest for ultimate truth stays focused on those theoretical dimensions inaccessible to our unaided senses; as a result, more and more of us come to assume that those theoretical realms are more true, more fundamental, more *real* than this palpable world that we experience with our breathing bodies.

But this assumption—this tendency to explain the lived world by reference to dimensions hidden beneath or beyond that world—actually inverts the demonstrable state of affairs. Not even our most brilliant researchers have continuous experience of those other dimensions; indeed, the empirical evidence regarding those realms is necessarily fragmentary and piecemeal. Science encounters those alien scales of existence only by means of exceedingly specialized instruments, and often only for brief moments—usually during highly structured experiments—when particular reactions are registered by such instruments, or specific measurements are recorded by means of them. Our scientists, then, must fill in the yawning gaps between those fragmentary glimpses in order to glean some sense of how those hidden worlds really function—how "quarks" are held within a proton or neutron; how particular stretches of DNA influence other stretches of the same chromosome; how the neural lattice of branching axons and dendrites continually interacts with itself; how black holes twist the fabric of space-time. Our experts construct a hypothetical image of what's unfolding at those other scales only by extrapolating from their fragmentary findings, and creatively filling in the holes between them. Yet the manner in which they fill those vast gaps in the empirical data is inevitably shaped by intuitions, expectations, proclivities, and perceptual habits *borrowed from their ongoing (and taken-for-granted) engagement with the one realm that they inhabit with the whole of their animal bodies*—from this enigmatic, earthly cosmos of ground, wind, and rain that commonly meets their senses as they daily go about their lives.

Every coherent image we can have of those other, ostensibly more objective dimensions of reality is secretly rooted, then, in the ambiguous, ever-shifting terrain of our ordinary experience. Everything we have come to believe about those presumably more fundamental scales of reality is tacitly dependent upon our everyday engagement with the world at *this* scale—the very scale of existence to which our animal senses are tuned. In the open expanse of our full-bodied experience, the entities that meet our senses are not quarks and protons, but rather brambles and mushrooms and slowly eroding hillsides; not DNA base pairs or neuronal synapses but rather children, and woodpeckers, and the distant sound of thunder. Although we regularly try to explain this ambiguous world by appealing to those more mathematically precise realms hidden behind the perceivable surroundings, it is ultimately our ongoing relation to the capricious earth that holds the key to all those abstract and provisional worlds.

❧

Let's reprise this reflection from another angle, leaning for a moment upon the work of one of the most lucid observers of the world that meets our unaided senses, a scientist who rarely let his ponderings swerve too far from the earth of his direct experience. For more than two millennia, alphabetic civilization had claimed a special origin for humankind, asserting that humans were divinely fashioned to serve as God's representatives on earth. This presumption was challenged and largely overturned by Charles Darwin, whose painstaking observations provided convincing evidence that humankind, on the contrary, had taken shape like other species over the long course of evolution. The theory of evolution by natural selection made evident that we, too, are animals, created not by an external divinity but evolved from a group of primate ancestors, and more anciently from a lineage of small mammals, themselves derived, much earlier, from an ancestral line of fish long immersed in the ocean depths.

Darwin had rediscovered the deep truth of *totemism*—the animistic assumption, common to countless indigenous cultures but long banished from polite society, that human beings are closely kindred to other creatures, and indeed have various other animals as our direct ancestors. Here was a form of totemism transposed into the modern world—the totemic insight now translated into the language of "descent by natural selection from a common ancestor." This modern version no longer saw different persons as descendants of *different* totemic animals, but recognized all humankind as derived from a common lineage of creatures. In the wake of Darwin's bold insights, we have learned to consider all humans as members of a common family. But the wild, animistic implication of Darwin's insight has taken much longer to surface in our collective awareness, no doubt because it greatly threatens our cherished belief in human transcendence. Nonetheless, it is an inescapable implication of the evolutionary insight: we humans are corporeally related, by direct and indirect webs of evolutionary affiliation, to *every* other organism that we encounter.

Moreover, it is not only other animals, plants, and simpler organisms that have contributed, during the course of evolution, to the unique character of the human creature, but also the fluid ocean, and the many rocks that compose the soils, and the way the mountains gather clouds above the high ridges. These planetary structures are not extrinsic to human life—*they are not arbitrary or random aspects of a world we just happen to inhabit.* Rather they are the constitutive powers that summoned us into existence, and hence are the secret allies, the totemic guides, of all our actions. They are as much within us as they are around us; they compose the wider, deeper life of which our bodies are a part.

If we accept Darwin's insights, and concede that the human species has been shaped by the creative flux of evolution, then we must acknowledge that the enfolding biosphere is the very matrix within which our organism came to acquire its current form. Our senses have coevolved with the chemistry of these waters and this air, shaping themselves to the particular patterns of the animate earth. Our human eyes have evolved in subtle interaction with other, non-human eyes—as our ears are now tuned, by their very structure, to the howling of wolves and the thrumming of frogs. While gliding in huge, undulant schools through the depths of the amniotic oceans, or later, while crawling upon our bellies from puddle to puddle (our scaly skins glinting in the sun)—while racing beneath the grasses as tiny, nocturnal mammals, or leaping from branch to branch as long-tailed primates—our brainy bodies have steadily formed themselves in dynamic interaction with the textures and rhythms of terrestrial nature.

Our nervous systems, then, are wholly informed by the particular gravity of this sphere, by the way the sun's light filters down through Earth's atmosphere, and by the cyclical tug of Earth's moon. In a thoroughly palpable sense, we are born of this planet, our attentive bodies coevolved in rich and intimate rapport with the other bodily forms—animals, plants, mountains, rivers—that compose the shifting flesh of this breathing world.

So it is the elemental Earth that has lent us our particular proclivities and gifts, our specific styles of behavior. Our ways of moving, our modes of perception, our unique habits of thought and contemplation, have all been informed by the variegated nature of this wild-flowering world. Thus, the enfolding biosphere provides the *inescapable template* for our experience of any other realm we may discover or devise. Whether we are pondering the unfathomably vast galactic clusters revealed by a new generation of radio telescopes, or mathematically exploring the submicroscopic realm of vibrating, supersymmetric strings, we cannot help but interpret whatever we glimpse of those worlds *according to predilections derived from the one realm in which we uninterruptedly live*—predispositions necessarily instilled, over the marathon course of evolution, by the shapes, the patterns, and the other inhabitants of our terrestrial environment.

For example, our surmises regarding the subtle function of neural processes within the brain are profoundly constrained by the fact that the brain did not evolve in order to understand itself. The complex organization of the brain evolved as a consequence of our sensorial and muscled engagement with the complex, dangerous, and ever-shifting landscapes that surrounded us. The brain has thus a natural proclivity to help us orient and interact with those enigmatic surroundings. Whenever we attempt to focus the thinking brain back upon itself—upon its own neural structure and functioning, or upon other dimensions similarly hidden from our common experience, whether subatomic or cosmological—it cannot help but bring those predispositions to bear, anticipating gravity, ground, and sky where they are not necessarily to be found, interpreting data according to the elemental constraints common to our two-legged species, yielding an image of things profoundly informed by our animal body and its accustomed habitat.

There is much to be gleaned from our investigations into other scales and dimensions, yet we constantly err by assuming our studies provide an objective assessment of the way those other scales *really are in themselves,* independent of the very partial perspective that we curious primates have on those dimensions. The supersmall and ultravast spaces steadily explored by the sciences can never *explain* this lovely and problem-ridden world that we daily inhabit, *since those abstract spaces are largely woven out of the perceptual fabric of this very world.* Certainly our forays into those abstract dimensions can offer us clues, new approaches to the land around us, new ways of looking at a forest or feeling a volcanic tremor. They can spark new insights into the lives that surround us, new ways of understanding ourselves, or—if we choose—of altering ourselves. Ultimately, however, it is only the lived, felt relationships that we daily maintain with one another, with the other creatures that surround us and the terrain that sustains us, that can teach us the use and misuse of all our abstractions.

Despite all our giddy technological dreams, this vast and inscrutable land—drenched by the rains and parched by the summer sun—remains the ultimate ground, and the final horizon, of all our science. It is not primarily a set of mechanisms waiting to be figured out, this breathing land. It is not a stock of resources waiting to be utilized by us, or a storehouse of raw materials waiting to be developed. It is not an object.

It is, rather, the very body of wonder—a shuddering field of intelligence in whose round life we participate. And if, today, this dreaming land has been forgotten behind a clutch of glowing screens that intercept the fascination of our focused eyes—if it has been eclipsed by styles of speaking that deaden our senses, and by machinic modes of activity that stifle the eros between our body and the leafing forests—then it is time to listen, underneath all these words, for the animal stirrings that move within our limbs and our swelling torsos. It is time to unplug our gaze from the humming screen, walking out of the house to blink and piss under the river of stars. There are new stories waiting in the cool grasses, and new songs . . .

DEPTH

(Depth Ecology II)

> I only went out for a walk, and finally concluded to stay out till sundown, for going out, I found, was really going in.
>
> —JOHN MUIR

A certain blue of the sky is so damn blue that only blood could be more red. Actually, the French poet Paul Claudel was speaking of the blue of the *sea* when he made that curious observation, but up at this altitude, the sky itself is the sea I'm swimming in. Relentless blue fills every hollow, snuggling up against the outline of each boulder, drinking aspen sap straight from the bruised bark of these groves, pressing in on my eyes, rounding my nostrils and seeping in through my sweating pores as I walk down this slope. Maybe I'd better say that I'm loping rather than walking, since gravity's granting a long, fluid stride to my legs that they never have on level ground. Spruces and firs and clustered aspens glide past me as I descend, all heading in the other direction. But I pay them scant mind; I'm watching the shifting dance of the hills just in front of me.

Only a short time before, I had gotten to my feet, shoved my journal into my pack, taken a last look around the summit, and began hiking down this grassy incline. I couldn't see those dancing hills as yet, or rather I didn't notice them; they seemed part of the flat and forested valley floor that stretches westward from the foot of the mountain.

I had no map with me. I hadn't wanted to spend the requisite eighteen bucks for a decent topo map of these mountains, since I didn't really know what stretch of this terrain I'd choose to ramble. Then my car broke down. Impatient to get off the dirt road and into the woods, I took the breakdown as a sign to head in from that spot and see where I wound up.

It was a stupid move, but the exhausted old Ford didn't respond to my clumsy tinkering, and I didn't want to burn up the whole day trying to find a more reputable mechanic. Besides, I generally enjoy being lost—it being the quickest way I know to rouse my creaturely senses from their slumber and coax an entry into that elixir-like state of mind called "wilderness." Sure enough, that wild flavor had come on as soon as I'd wandered far enough to feel unsure just which ravine led back to my vehicle. By midafternoon I'd already hiked up and over several ridges; upon hauling my limbs to the top of the last modest peak, I could see that it was the westernmost summit in this part of the mountains. From that height, the western slope seemed to drop precipitously to the broad plain far below; I predicted that I'd soon be strolling six or seven level miles to a paved highway whose sinuous line I could just barely make out from the top.

But not five minutes after I began my loping descent, that valley floor extending so serene and flat toward the western horizon began undulating and folding. As though prompted by the intensity of my gaze as I scanned its forests for the best route to take, the valley began to lift various parts of itself skyward, like large swells rising from the green surface of a sea. The lower I got, the higher those waves ascended, the furrows between them deepening, until I realized that I was descending into a rolling maze of hills and ravines from which I was unlikely to emerge before dark. A mix of trepidation and elation had then spread into my muscles. I still had a fair amount of water, and although I'd brought no sleeping bag on this jaunt, I had a down jacket stuffed in my pack. The moon had been full two nights back, so if the sky stayed free of clouds I'd be able to hike through much of the night and catch a ride on that highway come morning.

I found myself grinning at the way I'd been fooled by the topography. My eyes seemed keener now, and the blue of the sky more penetrating; the apparent metamorphosis in the land had induced a new lucidity within my organism. And it was then, in midstride, that I noticed the hills slowly dancing.

Look: that steep slope across the way is even now splitting into two thickly wooded slopes, one gliding toward me and the other gliding backward, as a wide swale—invisible a moment ago—is opening between. From within that hollow, or rather through it, a distant ocher cliff is now beckoning. So this landscape is still metamorphosing; it didn't halt its transformations after these foothills emerged from what had seemed a flat plane—now the hills are shifting among themselves as I walk. They alter their relation to one another, some stepping to the fore while others hide themselves (sometimes reappearing a minute later from a new angle, transfigured and tough to recognize). That ocher-gray cliff that called out to me a moment ago has been swallowed up once again by this protean terrain, while a wooded hillside is now extending a muscled shoulder in my direction, reaching its arm out to hold and merge with the steep slope that I'm descending. A few conifers leaning out from that ridge coax me toward it. It appears much too narrow to safely cross, but as I approach it the ridge thickens and widens out; there's a faint path worn by hooves along the top edge. A horde of crows flaps overhead, their raucous cries filling my awareness until they blessedly dissolve back into the blue silence. Let's follow this deer trail and see if it offers a way though these labyrinthine hills.

I step onto the ridge and make my way along it, aided by random tree limbs offering handholds, then walk into the thicker woods on the opposite slope.

Depth is the dimension of closeness and distance. It is the way the world spreads out from a few aspen leaves left dangling, golden and spent, on a branch in front of your face, to the rocky ridge across the glen, and beyond that to the further hills piled one upon the other against the last mountain range silhouetted zigzag against the sky, and on past, opening out across the unseen desert to a horizon still faintly visible between a few peaks in that range. Depth is the visceral stretch between the near and far of things—the continuum, or glide, between the known and the unknown. It is the manner in which the distances beyond the horizon—realms that you can dream but cannot see—somehow open onto those vague and far-off shapes that you *can* see but cannot possibly touch, the way the perceptual world bounded by those edges swerves toward you out of that purely visible distance, growing more and more palpable until, yes, you can also reach it with your fingers, feeling the smooth texture of the branch with your hands as well as your eyes, a dimension that keeps on coming right up to your face and includes it.

Depth implicates the whole of our animal body (this carnal density of muscles and skin and breath), situating us physically within the animate landscape. Whenever we notice that some aspects of the perceived world are closer and more accessible to us than others, whenever we acknowledge that some phenomena are crisply visible while others are concealed from view, we affirm our bodily location *in the midst* of those phenomena. Unlike the dimensions of height and of width, which seem entirely objective attributes of the things that we perceive, the dimension of depth varies with each change in our own position. The overall *height* of the aspen grove across the meadow is a fairly fixed property of those clustered trees—as even the *width* of that grove seems an objective property of that place. But the *depth* of that grove manifestly depends upon my bodily position in relation to those trees. From where I stand, the aspen grove has a somewhat shallow depth; my gaze can readily pass through it to the darker evergreens on the far side. Yet as I walk around the grove, the depth shifts: from here, that crowd of aspens has a more profound and many-layered depth, and my eyes cannot really penetrate it. And if I step among those aspens, into the heart of the grove, then its depth will open and envelop me, and I will need to describe it afresh.

Check this, reader, against your own experience: unlike the height of a mountain range, and the width or span of a valley, the depth of a terrain—the relation between the near and far aspects of that land—depends entirely upon where you are standing *within* that terrain. As you move, bodily, within that landscape, the depth of the scape alters around you.

Such alterations of perceived depth are provoked not only by our own changes of position; they can also be triggered by activity in the land itself. When I am exploring an unfamiliar locale and find myself pondering the contour of a nearby hill, it may happen that a flock of chattering birds passes overhead, flying in the direction of that hillside, and I think that at any moment they are about to alight upon it—yet they keep flying, seeming to rise above the crest of that hill and then passing beyond it, becoming smaller and smaller in the sky, their calls growing fainter and fainter. Until, now barely visible, they abruptly dip down and alight, to my startlement, upon that very hillside! My eyes blink and stare: those birds had not flown past that hill at all—rather, that slope is much *farther away* than I'd thought, and so it has taken the flock all this time to reach it. Which necessarily means that that hill is a great deal larger than I'd assumed! Yet these thoughts are not undertaken consciously; rather, my animal body, thrown out of equilibrium by the incongruous landing of that distant flock, rapidly recalibrates its sensations, allowing a new experience of that hill to settle into my eyes and muscles, allowing the depth of this landscape to dilate and expand around me.

At yet another time, when walking along a slender stream in the high desert backcountry, our legs will carry us beneath a tall bank. And so the broad plain on which we'd been walking, dotted with cacti and juniper bushes, vanishes entirely from view; we're now stumbling within a claustrophobic realm, a world shrunken to the narrow confines of this creek's shouldered passage. Yet as we follow the creek upstream, stepping across it when we need to, this narrow world slowly begins to divulge its myriad secrets—its hidden nooks piled with patterned stones, or strewn with translucent exoskeletons shed by some sort of insect: here a fox's tracks pressed in the mud where she came to drink, there a bright lizard doing push-ups on a lichened rock—so that soon enough we are taken up within the strange interiority of this canyon cosmos, our senses tuning themselves to the manifold intricacies of a depth that keeps expanding. Hence we are taken aback when the canyon walls suddenly dwindle and our legs abruptly emerge into the wider landscape; we feel at first all exposed, visible, vulnerable on every side. Our eyes gradually stretch open their focus to meet this wider depth as our stride lengthens to walk within it. Meanwhile, the unique interiority of the canyon, its intricate and intimate immensity, begins to slip away from our awareness. In truth, we think now, it was less a canyon than a gully, and indeed when we glance back toward it as we hike away from the stream we now see only a linear interruption in the flat surface of the desert, like the top of a small, winding ditch, with no hint of the cosmos hidden within it.

Or at dusk in the thick of the city, walking through a different kind of canyon fashioned of gleaming marble and glass mirroring countless other pedestrians, reflecting yellow taxis and shiny limousines and tall buses loudly wheezing as they stop and start up again—vertical vaults of stone and glass reflecting, too, the screech of tires and the rustling of bags and the click of heels on the endless pavement and the intermittent honking of the cabs, echoing even your own voice as you yell "I love you, too!" into a cell phone held up to your ear and then snapped shut as you plunge forward, utterly focused on your destination off ahead in the rectilinear distance, when suddenly your ears clue you in to an eerie spaciousness to your right as the echoing wall you've been walking past is no longer there. You've reached the corner, and as you step down from the curb you swivel your head to glance up the cross street and there, smack-dab in the middle of that side canyon, is the full moon, swollen, rising above the tops of the cars, muffling the honking horns and turning a hundred panes of glass on both sides of the canyon into dazzling white facets, stealing your breath and crumpling your knees at the gorgeousness of a blaring city transformed into a silent, luminous crystal...

❧

Many years ago I journeyed through Southeast Asia, making my way as an itinerant sleight-of-hand magician, performing on street corners and at small family inns. One afternoon, I found myself walking alone between two tiny hamlets high in the Nepali Himalayas, under a towering mountain called Ama Dablam by the Sherpa villagers. All day long I'd been gazing the resplendent southwest visage of this mountain, its features etched with snow and ice, watching its expressions shift with the angle of the sun as I hiked up the narrow trail. In midafternoon I took off my backpack to rest for a moment, gulping some water and glancing behind me. I saw then that the monsoon clouds were slowly ascending the southern valleys at my back. I hoisted my pack and quickened my pace, with the intent to reach my destination—the isolated home of a Sherpa *jhankri,* or shaman—while the visibility was still good. But one appendage of that cloud mass seemed to advance more rapidly than the rest, climbing the slopes at the base of Ama Dablam; soon it was caressing the sharp edge of the mountain wall. I stopped to watch, bowing low to the mountain, thanking it for the indulgence of its companionship and for its protection (Ama means "mother" in Sherpa, as it does in many other languages) as the cloud slowly came between Ama Dablam and my eyes. For a while I caught glimpses of the mountain's ravishing face through various rents and rifts in the cloud, but then the white mass thickened, and I could no longer see it. Ama Dablam was gone. Trying to feel its implacable presence behind the gaseous miasma, I was astonished to discover that I could no longer sense the mountain at all. I was no more able to *feel* the presence of that looming peak, now, than I was able to *see* it. The mountain, whose majestic bulk had commanded my attention for most of the day, had simply vanished.

It was the purest disappearance I'd ever witnessed—no doubt because the entity who was suddenly *not there* was a being of such immutable power, a great wave of earth whose adamant eloquence had been conversing with my sweating body for so many strenuous hours. It occurred to me then, staring across the boulder-strewn pass into the swirling vapor, that here was the truest archetype for my craft as a sleight-of-hand magician. I must strive to make my silver coins and my billiard balls vanish as completely as this cloud had coaxed the entire mountain to vanish, transforming an immense colossus of rock and ice into a fleeting phantom, a mere memory. To this day, whenever I roll a silver dollar across my knuckles, preparing to make it disappear, I bow subtly toward the remembered mountain and the monsoon clouds—an apprentice honoring his masters.

Sleight-of-hand relies entirely on the dimension of depth, since this dimension alone empowers nearby surfaces to hide farther ones, enabling the back of a relaxed hand to conceal a ball gently gripped in its palm. The same depth that empowers clouds to dissolve cliffs also allows the dangling folds of a previously empty scarf to conceal a patient rabbit held behind them. Yet depth can also work its magic entirely out in the open. While strolling barefoot along a beach you might glimpse, in the near distance, a discarded tire half-buried in the sand, its worn-out treads exposed to the sun and salt spray. If you stroll closer, however, you may see that tire suddenly metamorphose into a dozing seal that rouses and barks, gruffly, at your approach, and then galumphs into the water.

Some months before my extended trek in the Himalayas, I was exploring a wild peninsula jutting into the Indian Ocean from the south coast of Java. It was a sweltering hot day, with no breeze to ease the thickness; I had peeled off my sweat-soaked shirt and tied it 'round my waist. Emerging from some woods into a wide field above the sea, I saw that a wind was jostling the leaves and swaying the branches of some trees on the far side of the clearing. Grateful, I made my way toward that grove in order to feel the moving air on my damp skin. As I got close to those trees, however, the apparent wind bending their branches abruptly transformed into a band of monkeys foraging for food among their leaves. There was no breeze at all, only a bunch of primates better adapted to the heat than I.

Such perceptual transformations are endemic to a reality that exists only in depth—a reality that discloses itself to us only by holding some part of itself in the uncertain distance, a world that we encounter only in the tension between the nearby and the yonder. The dimension of depth is ruled by Proteus, the shapeshifting god, whose dominion over sensory experience teaches us to stay loose, to invite magic, to expect metamorphosis. While lying belly-down on the grass one afternoon, my head propped on one hand as the other turned the page of a novel, I took vague note of an airplane soaring far above at the very periphery of my vision, then became alarmed when its engine grew louder and I sensed it beginning a nosedive. Only as I frantically twisted my torso to stare did the careening plane suddenly transform, resolving into a buzzing bottle fly walking down the right lens of my eyeglasses.

Of course, such perceptual shifts, slippages, and corrections are occurring all the time as we go about our days. If we rarely notice such transformations today—if we seem to have lost our sense of the earth's audacious and metamorphic magic—it is likely because our depth perception has become impoverished. After all, our sensorial engagement with the ambiguous depth of our world has been largely overcome, in the last half century, by our steady involvement with flat *representations* of that world. So many children have been raised, day after day, on the flat screens of televisions, video monitors, and handheld computers—as indeed many adults spend their waking hours focused upon digital screens! This trading of the world's visceral ambiguity for flat representations is a habit that began centuries ago, when cultural elites began to devalue the participatory wisdom drawn directly from physical engagement with the living land in favor of knowledge drawn, secondhand, from the flat pages of books. It matters little that the things written of on those pages may be filled with creative nuance, or that the glowing screen carries an image rich with perspectival programming and simulated depth—for it is first the flat surface that intercedes between us and that depth. Our animal senses are no longer in direct relation with the sensuous terrain; our muscled body sits immobilized before the smooth and scintillating surface upon which we gaze, enthralled.

My laptop computer has a built-in DVD player. And so, for once, I am watching a film on my laptop. At a crucial point in the story, a dilapidated truck appears and parks in the background, off behind the main character. Yet my eyes do not alter the depth at which they're focused in order to clarify that distant truck. For the camera filming the scene has already done the focusing for me. As I glance toward that truck, my eyes remain focused at the very same distance as when they were watching the central character in the foreground—a distance that, in truth, is not fifteen feet away from me, where that fellow *appears* to be standing, but merely the thirty-inch distance of the computer screen from my face. If the artful cameraman had left that far-off truck out of focus, then my visual system would be unable to bring it into focus in any case—whether by tightening the tiny muscles around the retina of each eye, or by changing the angle at which my two eyes converge. Having watched many films, my eyes long ago gave up trying to focus the individual images; they simply surrender themselves to the screen.

As we spend more and more of our lives gazing at screens, tuning our senses to the fabricated distances that we see there, the instinctive participation of our eyes in the near and far of the world is suspended for longer and longer periods. The ability to actively engage the manifold and shifting depth of the land—altering one's focus from a close clump of grass to the far-off cliffs, then abruptly back to fix on a woodpecker thwacking a dead snag in the intermediate distance—disarticulates and dissolves in the face of this new entrancement with a world displayed at an exact remove from one's face.

If you are watching a nature-based program on television—observing a female lion, perhaps, as she lolls with her cubs under the shade of an acacia tree—and you happen to stand up and walk across the room, notice: your movement does not alter anything on the screen. The depth of the room, of course, shifts around you as you move—the bookcase looms up in front of you and then recedes as you step past it, the music stand momentarily comes between you and the television as you walk by—yet the spatial positions of those cubs do not shift in relation to one another, or to the acacia tree behind them. For you and those lions do not inhabit the same space. There is no *depth* between you and those creatures, since you stare at them from a position entirely outside of their world.

You may, of course, learn a great deal about lions by watching such a program. But the primary lesson your organism steadily learns from the program is that nature is something you look *at,* not something you are *in* and *of.*

Sometimes the screens that hold our gaze contain no images at all, only numerical equations, or lists, or carefully crafted sentences. Back at my laptop now, I sit gazing at the white expanse, tapping with my fingers as letters array themselves on its surface like ants. It is a compelling enough activity, watching the lines of ants assemble, although when I raise my eyes to glance out the window, I do not see the many-branched cottonwood in its inviting depth, nor sense its near branches pressing against the pane. I do not feel the distance between this cabin and the twin hills across the valley, although they, too, are visible through the glass. I don't sense the heavy bulk of those hills, or the relative lightness of the clouds clumped above them. For my screen-dazzled eyes do not focus upon the swooping branches or those far hills; they see only these blurred and somewhat flattened shapes juxtaposed as on a single plane—forming, as it were, a kind of painted backdrop for my cogitations. What I experience, in other words, is neither that cottonwood nor those clouds, but simply "the scenery." Only when I take a break from the writing and launch myself from the desk, walking to the door and yanking it open, the cold air shaking my brain loose from its verbal exertions—only then does that scenery begin to breathe, the barks of sundry dogs stretching open the visible landscape as their contrasting nearness or distance registers in my ears, and the hills stand up one in front of the other, and branches dangle near my head—and so my body finally remembers itself back inside the round life of the local earth.

The more we spend our days staring at screens, taking our dreams and directives from the signs and shapes that play across their smooth surfaces, the harder it becomes to make this transition, the easier to simply stay ensconced within a universe of images neatly honed to meet our needs. Even if we venture beyond the walls of our office or metropolis, we often find ourselves merely staring at the scenery. Accustomed to peering at flat representations, we've begun to take the palpable world itself as a kind of representation—no longer a limitless field in whose boisterous life we're participant, but a set of determinate facts arrayed in front of us, upon which we gaze like detached and impartial spectators. We no longer peer into the enigmatic depths of a terrain that encompasses and exceeds us; the land has now become something that we look *at.*

For example, today many educated folks have ceased experiencing the earth as the rough ground that supports their steps and the wind pouring past their ears; for them the earth has become, first and foremost, that blue sphere flecked with clouds that they have seen in various NASA photos—images that were reproduced countless times and became the basis for a thousand corporate logos. For a great many persons the earth is now equated, in other words, with the earth *seen from outside*—with our planet as it might be seen by a transcendent god, or by a surveillance satellite. This "objective" view of the earth now conditions and supersedes our immediate experience of the rainswept land that enfolds us and drenches our shirt as we run to the mailbox. Such soaked and shivering experiences, we now realize, are secondary, mere "subjective" distortions of a factual reality that has already been mapped, measured, and monitored. It may indeed be raining, but we can best clarify this by turning on the Weather Channel.

To our animal awareness the land remains ambiguous, open-ended, and always prone (as we saw earlier) to unexpected metamorphoses. But such visceral experiences seem arbitrary and ephemeral next to the "hard" facts yielded by the impersonal instruments now surveying the real. Most of us moderns pay scant attention to our directly felt impressions of the world. We hardly notice such impressions anymore, or we straightaway translate those qualitative sensations into the quantitative world of facts, and information, that is the primary object of faith in our era.

It is a faith partly held in place, I'm suggesting, by our ever-growing participation with screens of all kinds, by our steady subjection to flat representations and spectacles. The belief in a thoroughly objective comprehension of nature, our aspiration to a clear and complete understanding of how the world works, is precisely the belief in an entirely *flat* world seen from above, a world without depth, a nature that we are not a part of but that we stare at from outside—like a disembodied mind, or like a person gazing at a computer model.

When we consider the things from a disincarnate position outside the world that they inhabit, the ambiguous effects of depth dissipate: we seem to encounter the things as they really are, naked and exposed to an all-seeing awareness that, freed from any limited perspective, is able to penetrate every opacity, able to register every aspect of the real. By renouncing our animal embodiment—by pretending to be disembodied minds looking *at* nature without being *situated* within it—we dispel the tricksterlike magic of the world. We dissipate its mysterious concealments and transfigurations, dissolving at last the confounding limitations that the dimension of depth had imposed upon all our knowledge.

And yet it remains a pretense. Whenever we view the natural world as a set of objects, whenever we perceive the terrain around us as a clear and determinable set of facts, then we are in truth not really perceiving the actual terrain at all, but only a representation—a projected concept—of that terrain. We may *conceive* of earthly reality as though we were not ourselves of it, but we can never really *perceive* it as such.

Although he lusted for the hard matter of reality, Henry Thoreau, exemplary animal that he was, knew well that the living, breathing body could never encounter a wholly determinate and defined object, and hence that we never really meet a naked fact:

> If you stand right fronting and face-to-face to a fact, you will see the sun glimmer on both its surfaces, as if it were a scimitar, and feel its sweet edge dividing you through the heart and marrow, and so you will happily conclude your mortal career.[Ω]

Yet we've convinced ourselves that the land around us—generously abundant at some moments, ferocious and unforgiving at others—is really a set of sheer facts; we hold ourselves apart from the world in order to subdue its wildness.

The mechanical and mindless nature that we ponder under the cool guise of objectivity has proved outrageously useful, a model that has enabled the most audacious manipulations undertaken in the name of betterment and progress, from the diversion of rivers and their transformation into electricity, to the mining of genomes and their transformation into prestige and profit. Moreover, it's now clear that the objective model of nature will be necessary for the careful mending of the many world-wounds brought about by all this progress. We will not get by without it. Yet it is a model nonetheless, a flattened representation of nature. It is not the wild world in its living actuality—not this mystery that we contact with our own breathing and bleeding bodies. It is not the cottonwood branch now tangled in my windblown hair, or the flock of cranes swirling far overhead in midmigration, their ratchetlike croaks filtering down through the tumult of sky. Strand by strand, I ease my hair free from the branch. Near the base of the tree, a small stone

[Ω] Henry David Thoreau, *Walden* (New York: Longmans, Green, and Co., 1910), p. 82.

glints silver against the dark soil. It is not that stone, nor the sunlight on that stone.

The density of these woods must be muffling the soundscape—I've heard neither birds nor the wind for a long time, only the sound of my footfalls as they crunch along the deer trail. The sun is much lower than when I first noticed the hills dancing before me; I'm in the thick of those hills now, following this faint trail strewn with needles among the tall pines and firs. It's led me over large outcroppings and into rock-strewn arroyos, where I commonly lose the path and have to wander back and forth on the other side of the ravine hoping to find it again. The late afternoon shadows are what throw me off, disclosing apparent passages through the undergrowth that dead-end as soon as I try them. Only by blundering up and down do I happen upon the right angle from which to glimpse the actual way through, and so my feet finally rejoin the crowd of hooves and paws that have passed along this route.

This animal road, slight as it is, is more inviting than the alternative—following one of the ravines down to where it would join a larger streambed. These steep gashes are too cluttered with jutting roots and tumbled rocks to allow ready passage. So I keep cutting across the ravines, wending my way laterally along these wooded slopes, trusting the trail. I know I'm walking vaguely westward since the sun glimmers and glints through the canopy in front of me, its long rays slanting though the trees. Thin trunks and thick—saplings and elders, and middle-aged trunks as well—all glide past me as I walk, though at different speeds. Those nearest to the trail rush past, their limbs brushing my shoulders or slapping my legs, while those a few yards off move with less haste. Farther still from the trail are trees whose gait is leisurely, even stately. And beyond even those are the grand but distant trees that seem to linger with me as I stride; now and then one of them catches and holds my eye—I glimpse it again and yet again through the rapid syncopation of nearby trunks.

Farther off to my left, darkly green behind all of these gliding verticals, are the needled trees that blanket the hillside across from this one. The forest on that facing slope appears to keep pace with me as I stroll, while all these nearer conifers slide past at their different rates. It's as though all the trees on that hillside were gliding forward—or is it just the contrast with the nearby woods rushing backward that makes the distant slope seem to stride in the opposite direction?

But wait! What am I saying?! None of these trees is actually moving! I alone, among all these upright bodies, am actually locomoting along the ground. All their apparent movements, fast and slow, are merely a consequence of my own physical motion, the illusory effect of my own activity in the midst of what is, in fact, a fairly quiescent and passive topography. "Motion parallax" is the technical term for this dynamism that my own movement seems to induce in the landscape, this apparent roaming of things in relation to myself as I wander.

As soon as I reflect upon this, the land's activity seems to subside. A certain vitality I had sensed in the forest has now dissolved. The dynamism has withdrawn from the surroundings and has concentrated itself within my skull, where all these thoughts are now churning (calculating the way that these hills and carved ravines must look to a detached eye situated above the landscape, pondering how that more objective view is distorted when the terrain is seen from a particular position within it, moving at this particular pace). I am lost in my figurations. My bodily senses blunted, the land around me has become a passive field of shapes, some of them neatly defined ("Douglas fir," "blue spruce"), others obscure and unnoticed as my thinking self plunges past. I am now staring at my watch, worrying about my broken-down car. I am wondering when I might emerge from these wooded valleys and find my way to the highway . . .

A spiderweb collapses across my face, sticky and strange, like a Bazooka bubble-gum bubble that I once blew to bursting and had to peel, paper-thin, from my cheeks. I wipe the web from my nose and from the edge of my baseball cap; but before I can place the cap back on my head I notice the spider, so close I have to cross my eyes to see her brown body clearly, suspended by a single strand from the beak of my cap. Dangling upside down with her legs folded inward, the spider is falling away from the cap on the rapidly lengthening span. "Hello there," I say hopefully, raising the baseball cap high so the spider dangles again at eye level, but she keeps riding the end of that thread down and away. I carry the cap over to a nearby scrub oak, and transfer the cap's end of the thread to an upper leaf. The veins in those leaves are clear and crisply articulated to my gaze, as are the tiny nodules toward the end of the branch where next year's buds are already preparing themselves. By holding my eyes for a few moments, the spider has drawn my senses into another scale of experience. Which opens up endlessly. On a neighboring bush the leaves are different, the lobes rounder; here several branches have had their tips broken, or rather—given the frayed ends—chewed off. Deer, I guess, passing by in the early summer when the shoots were still green. I look down on the ground, brushing aside the fallen needles to glimpse some dried-out pellets, but find none—only a line of black ants in momentary disarray as a result of my sweeping. Three of them are scrambling, their antennae taut and probing, to find again the scent. Which they do straightaway, two of them bumping into each other and brushing their antennae together, briefly, before rejoining the line.

I continue along my own path, following not a scent but a vague absence of obstacles, the path more a visual hunch at this point than a clear passage, although as I walk it becomes more obvious, senses tuning themselves to the trail. Soon my peripheral awareness takes over the task of keeping me on track, and my gaze opens outward to touch the needled shadows and the sky above the balconied branches and the slope glimmering through the trunks to my left. That neighboring hillside, steeper now, is again gliding forward, accompanying me as I tramp, while between here and there the trunks move backward in their respective gaits, the farther ones slower and the close ones swifter, the nearest grabbing at my jacket as they rush past, and I realize that all this motion is not at all an illusion: it is the way the world lives, the way the world shows itself to itself—for there is NO view from outside! The layered dynamism of all these gliding trajectories has overcome a threshold in my self; my thoughts dissolve into my breathing body and I awaken as this striding form, this sentient flesh utterly immersed in the sensuous.

Nostrils flaring, arms swooping, I feel my animal solidity and grace as I lope down the trail.

The way that all these other bodies—trees, bushes, hillsides—shift in relation to one another as I walk *compels* my thorough inclusion in the landscape; when I really notice and pay attention to their transformations, I'm forced to discover myself utterly inside the physical world. These shifting gradients and angles of alignment—the multiple tree-lined corridors that seem to open around me as I move—all converge and cross here, at this animate creature that is me, ambling through this forest. There really *is* this huge world going about its business independent of me, and yet I am *in* it, alive in its folds!

High above me, the upper branches jostle and bend in gusts that I can't feel way down here, while the massive bulk of these hills shifts around me—I feel my own smallness among them. Yet no matter how minuscule, I also feel my own agency, my own autonomy within this massively real world. I'm embedded in this world, yes—but I am not bound, not imprisoned. There's a new freedom I feel, a looseness, an improvisational openness between myself and the beings around me. As if only by dissolving the distance between my mind and my body is this other, richer distance able to make itself felt—the open stretch between me and a far-off boulder, or the tension between my hand and the black scar burnt by lightning into the bark of a passing ponderosa. There's a dynamism to these distances, a palpable magnetism between my torso and that steep slope over there—an allurement made possible by the distance between things. Depth is not a determinate relation between inert objects arrayed within a static space, but a dynamic tension between bodies, between beings that beckon and repulse one another across an expanse that can never be precisely mapped—across a gap that waxes and wanes according to the mood playing through one's limbs or the limbs of the forest itself, according to the whoosh of the wind pouring through the needles, or the way the sun spills its warmth upon the soil.

Or now: the way that sun is sliding behind the trees in front, casting these woods into a denser weave of shadow, its rays no longer filtering down like white gold through the needled roof. The mood of this place alters, gently. A red squirrel is chittering off to my right, distressed by something. Perhaps by me. Animals less at ease in the daytime are also stirring throughout the forest. I cannot see them, yet there's a new wakefulness to the woods that sparks along the surface of my skin.

As the squirrel's staccato dies down, there remains a faint, high-pitched chirping reaching my ears from the upper canopy. Nuthatches? I'm not sure—the sound is so ephemeral, rising and fading . . . A jewel-like gleam flashes through the trunks in front of me, then vanishes, and then the sun flashes through again, steadying itself: the forest is thinning. Or rather, I am coming to an edge. As the last trees part around me, I step into a meadow of clumped grasses and wide swaths of barren rock. The breeze is stronger here, pouring upslope from the southwest as the low sun beats on my face. But it's the open sky above that halts my steps, that ocean of blue crashing down upon the rocks and the grasses, and upon my shoulders as well, scooping me up above the pinetops to soar and tumble in its deep expanse, then depositing me here where I stand, half floating, staring up.

The sky is not as clear as when I first stepped onto this deer trail two or three hours ago. A few clouds loaf overhead, their western edges gilded and glowing; others punctuate the blue toward the south. Watching their slow transformations in the silence of early evening, feeling the drift and loll of these porous, breathlike beings made of mist and yellow light, I realize the error of our common belief that we live *on* the earth. The rough-skinned rock beneath my feet is earth, yes, but what of those clouds, and the unseen sea in which those clouds are adrift? Are they not also part of earth? And if so, would it not be more true to say that we dwell *in* the earth, rather than on it?

The claim that we live "on earth's surface" implies that those clouds overhead are not themselves earthly powers, that the invisible depth in which they swim is an emptiness, a void continuous with the space between planets. If we dwell "on the earth," then those clouds are merely part of the flotsam and jetsam of space, with only a tenuous, neighborly link to the surface on which we stand. Yet this seems not quite right. We have heard that this planet is spinning like a dervish 'round its axis; here at this latitude the planet's surface is hurtling through space at around eight hundred miles per hour. We know that this rotation explains the apparent course of the sun across the sky, and the nightly passage of the stars through the darkness. Of course, standing here on the whirling earth I do not *seem* to be passing the sun with such swiftness; neither the sun nor the stars at night seem to move at anything approaching such a reckless speed! As I walked through the forest this afternoon, however, I saw that the trees farther from me moved past much slower than those close to the trail, and so it makes ready sense to me that the sun and the stars, which dwell so very far from this earth, appear to move rather leisurely across the heavens. But those clouds are much, much *closer* to where I stand than the sun or the other stars, just overhead really, not much higher than the mountains of this range. Rather like the near bushes and trees that sped by me, slapping my legs as I strode past, I should see those clouds hurtling westward as the earth's surface whips toward the east at breakneck speed. And that is the strangeness: those huge clouds are hardly moving at all! What's more, their lazy drift at this moment is not at all toward the west, but rather toward the northeast. Which confounds every notion of parallax, and confuses everything I saw today regarding the way the world moves in relation to me as I walk. If the earth's surface is really rushing eastward, then I should see those clouds racing rather recklessly toward the west!

Unless, of course, the clouds themselves *are a part of* the turning earth. In which case I am not really standing on the *surface* of this world, but am submerged within a transparent layer of this planet, an invisible stratum of the earth that extends far above those clouds. Unless, that is, the unseen substance that rides along these slopes, whispering the needles and rubbing the branches against one another—this tangible but utterly ungraspable element that swirls between my legs and pours in and out through my nostrils, the fluid medium through which those clouds are floating, and of whose lightness they partake, of which they are a kind of visible thickening or crystallization—is not at all continuous with the space that hosts the sun and stars, but is an element of this earth, an amniotic substance entirely proper to this place.

Which indeed it is. The air is not a random bunch of gases simply drawn to earth by the earth's gravity, but an elixir generated by the soils, the oceans, and the numberless organisms that inhabit this world, each creature exchanging certain ingredients for others as it inhales and exhales, drinking the sunlight with our leaves or filtering the water with our gills, all of us contributing to the composition of this phantasmagoric brew, circulating it steadily between us and nourishing ourselves on its magic, generating ourselves from its substance. It is as endemic to the earth as the sandstone beneath my boots. Perhaps we should add the letter *i* to our planet's name, and call it "Eairth," in order to remind ourselves that the "air" is entirely a part of the eairth, and the *i,* the I or self, is wholly immersed in that fluid element.

The gilt-edged clouds overhead are not plunging westward as the planet rolls beneath them because they themselves are a part of the rolling Eairth. Creatures of the embracing air, of an invisible but nonetheless material layer of this planet, the clouds accompany the Eairth as it turns, their shapeshifting bodies drifting this way and that with the winds. And we, imbibing and strolling through that same air, do not then live on the eairth but *in* it. We are enfolded within it, permeated, carnally immersed *in the depths* of this breathing planet.

MIND

(Knowledge II: The Ecology of Consciousness)

In the seventeenth century, as is well known, a certain French philosopher conceptually divided the world into two disparate substances: matter (*res extensa,* or "extended substance") and mind (*res cogitans,* or "thinking substance"). Matter, as René Descartes described it, was spatially extended, determinate, and mechanical; mind, on the contrary, had no spatial presence whatsoever—it was pure thought, free of all physical constraint and limitation. While animals, plants, and indeed nature as a whole were composed exclusively of mechanical matter, and while God consisted entirely of mind, humans alone—according to Descartes—were a mixture of the two substances. The human body, like other animal bodies, was a completely mechanical configuration; the immaterial mind somehow interacted with this physical body from a location within the human brain.

Descartes' resolute philosophical dualism neatly formalized a split between the mental and material domains that had long been implicit in European thought. In the centuries since, many theorists have attempted to reconcile these apparently contrary realms. Over the last hundred years, in particular, various natural scientists and philosophers have tried to reconstrue consciousness in purely material terms, speculating that the mind might be a product of the physical brain, or even identical to the material brain itself. The more subtle of these theorists suggested that mind should perhaps be understood less as a substance than as a material *process,* as the dynamic *activity* of the physical brain. Indeed, as the organization of the nervous system was gradually revealed over the course of the twentieth century, many scientists began to wonder if their own thoughts were nothing other than the felt experience of neuronal and neurochemical activites unfolding within their heads.

Some of these researchers suspected that they had finally overcome centuries of unnecessary dualism simply by doing away with the notion of an immaterial intelligence, or soul. Yet their new identification of mind with the dynamics of the physical brain quickly prompted another, more surreptitious dualism to spread throughout the secular culture of the West. This was the dualism between the brain—now considered the very organ of awareness—and the *rest* of the human body. No longer an opposition between an immaterial consciousness and a material flesh, the new, more subtle opposition was that between the sentient brain, conceived as a kind of magical mass of living matter, and a physical body still conceived in thoroughly mechanical and deterministic terms.

The tacit dichotomy between the mechanical body and the sentient brain was ultimately no less problematic than that between spirit and flesh; it left us unable to account for the interpenetration of these two very different types of matter. Nor was such a dichotomy warranted by the evidence. The brain can hardly be considered an autonomous organ neatly separable from the body, for the central nervous system is utterly integral to the whole body's functioning. It is difficult, if not impossible, to conceive of how a mindful brain could have arisen except as an attribute of a muscled and sensing organism—how a brain could evolve independent of a breathing body fending for itself in the biosphere.

Nor can the living body be understood in purely mechanical terms. Many of our unconcious bodily actions exhibit a creativity that cannot be accounted for on the basis of strictly mechanical physiology. The way our leg muscles, for example, dynamically adjust to the changing contour of a mountainside that we're descending, without our heady thoughts being at all cognizant of the process. Or the way my hand, interrupting its typing for a moment, swiftly navigates the distance between the keyboard and a bag on the shelf behind me, lifing a crisp corn chip to my lips before my conscious self is even aware of the action . . . Such unreflective yet remarkably attuned engagements of our limbs with the space around them suggest that even the most "automatic" actions of the body sometimes involve an open and improvisational intelligence that defies any neat distinction between a sentient brain on the one hand and an insentient or mindless physiology on the other.

Prompted by a growing awareness of such contradictions, in the final decades of the twentieth century a diverse group of theorists began to call attention to the profound importance of the body itself in the constitution of the mind. A small but influential array of researchers from various disciplines began to amass evidence demonstrating that mental experience was dependent not only upon the functioning brain but upon the whole of the animate organism—that the mind is less an attribute of the brain than of the living body as a whole, of which the brain is simply a necessary constituent.

Within the humanities, some of the key theorists drew inspiration from the writings published at midcentury by the French philosopher Maurice Merleau-Ponty, who had analyzed, with stunning lucidity, the body's influence upon even our most rarefied cogitiations. Others took their lead from the deepening feminist movement, with its incisive critique of patriarchal modes of thought; they questioned the privilege accorded to abstract, disembodied styles of reflection and began to disclose the hidden, overlooked intelligence of the human body itself, ascribing new value to corporeal forms of knowing. In the sciences, meanwhile, congruent research began to arise from diverse and divergent investigations. Biologist Francisco Varela and his colleagues employed their experiments to show that perceptual experience is thoroughly conditioned by the ongoing, active self-organization of the body as a whole. Cognitive scientists like George Lakoff and Mark

Johnson traced the development of mental categories and concepts back to their origins in the most basic and taken-for-granted bodily experiences. In the field of medicine, neurophysiologist Antonio Damasio demonstrated the profound dependence of rational thought upon the human body's more immediate emotions and feelings. According to Damasio, the mind, at its base, is nothing other than the body's ongoing experience of perturbations unfolding at the contact surface between itself and the world.[Ω]

Numerous other theorists could be named. Certainly the assorted proponents of this recent turn toward embodiment have often seemed radical, even iconoclastic, to the more conventional researchers around them. Yet in many ways their claims have simply resuscitated a long-overlooked, centuries-old approach to the mind-body problem—an approach developed in the seventeenth century by a young lens-grinder and philosopher named Baruch de Spinoza.

Born in 1632, Spinoza grew up in the large Sephardic Jewish community of seventeenth-century Amsterdam. The community was made up largely of *marranos,* Iberian Jews who'd been forced to convert to Christianity under the Inquisition, yet had continued to practice their faith in secret. Many marranos had recently fled Portugal for Amsterdam, at that time the most religiously tolerant

[Ω] See, for example, Maurice Merleau-Ponty, *Phenomenology of Perception,* trans. Colin Smith (London: Routledge & Kegan Paul, 1962), and *The Visible and the Invisible,* trans. Alphonso Lingis (Evanston, Ill.: Northwestern University Press, 1968); Janet Price and Margrit Shildrick, eds., *Feminist Theory and the Body: A Reader* (London: Routledge, 1999); Tamsin Lorraine, *Irigaray and Deleuze: Experiments in Visceral Philosophy* (Ithaca, N.Y.: Cornell University Press, 1999); Francisco Varela and Humberto Maturana, *Autopoiesis and Cognition: The Realization of the Living* (Boston: Reidel, 1980); Francisco Varela, Evan Thompson, and Eleanor Rosch, *The Embodied Mind: Cognitive Science and Human Experience* (Cambridge, Mass.: MIT Press, 1991); George Lakoff and Mark Johnson, *Metaphors We Live By* (Chicago: University of Chicago Press, 1980), and *Philosophy in the Flesh: The Embodied Mind and Its Challenge to Western Thought* (New York: Basic Books, 1999); Antonio Damasio, *Descartes' Error: Emotion, Reason, and the Human Brain* (New York: Putnam, 1994), and *The Feeling of What Happens: Body and Emotion in the Making of Consciousness* (New York: Harcourt, 2000).

city in Europe. As a teenager Spinoza, already a brilliant scholar of the Jewish textual tradition, began to question the received beliefs of the rabbis and other elders in the community. When he came upon the philosophical writings of his older contemporary, Descartes, Spinoza was taken by the carefully reasoned form of the French thinker's argumentation. Impressed by the clarity of the analyses, Spinoza nonetheless found himself dismayed by inconsistencies in Descartes' conclusions. Soon he set about adapting Descartes' rational style to his own, more subtle purposes.

Spinoza challenged the older philosopher's segregation of mental substance from material substance, arguing instead that mind and matter were not two separable substances but simply two different attributes, or aspects, of one and the same substance, which he called *Deus, sive Natura,* "God, or Nature." This unitary substance could appear either as matter, on the one hand, or as mind, on the other, depending upon the vantage we viewed it from. Just as, according to Spinoza, the vast and originating power that his contemporaries called "God" was nothing other than the creative dynamism and intelligence of Nature itself, so the human mind was simply the specific sensitivity and sentience of that part of nature we recognize as a human body. *Every* material body or thing, for Spinoza, had its mental aspect—all things were ensouled. The human body was the outward, material aspect of the human mind, as the mind was nothing other than the internal, felt experience of the body. "The mind and the body are one and the same thing . . ."[n]

It was such heretical assertions, articulated in numerous conversations with his contemporaries, that in his twenty-fourth year earned Spinoza the harshest possible reproach from the elders of the flourishing synagogue in Amsterdam: he was excommunicated, formally cursed, and banished from the Jewish community. Spinoza accepted this exile without the least objection, remarking only that it left him freer to pursue his researches without distrac-

[n] Spinoza, *Ethics,* Part Three, Proposition II.

tion. At once intensely rational and deeply spiritual, Spinoza was possessed of a remarkably empathic insight into the emotional lives of others. His radical identification of God with nature (and his insistence, starkly original at the time, upon the necessary separation of church and state) earned him an abundance of scorn throughout Christian Europe, ensuring that very little of his work could be published while he was alive. Yet such was the originality of his insights that some of the most innovative theorists of our own era claim Spinoza as their progenitor and inspiration.

Still, in one important respect Spinoza remains ahead of even those researchers who today claim his heretical insights as their own. Most of those who now assert the centrality of the body in any understanding of the mind—those who argue that it is really the body as a whole, and not an isolated brain, that is the true locus of awareness—still remain trapped within the confines of an unnecessary presumption. It's a presumption that lingers as our deepest inheritance from the Cartesian tradition: the assumption that awareness, or mind, is a special possession of our species, a property that isolates humankind from the rest of material nature.

The primary dichotomy in Descartes' philosophy, after all, was not the division between mind and body, but rather the divide between the mind and the *whole of the material world.* It was this more profound bifurcation between mind and matter that had ensured the rapid ascendancy of the Cartesian worldview and its long success in the West. By conceiving of itself as something entirely distinct from palpable nature, the rational mind of the Enlightenment was empowered to pursue its giddy dream of comprehending, and mastering, every aspect of the material cosmos. Descartes' segregation of the mind from the body, in other words, was but a means to a grander end; it authorized the modern mind to reflect upon the material world as though it were not a part of that world—to look upon nature from a cool, detached position ostensibly outside of that nature.

And so, today, if the long-held distinction between our minds and our bodies is disintegrating in the face of researches from a wide range of disciplines, we may wonder whether this other, deeper segregation—between mind and materiality, or between sentience and the sensuous cosmos—is itself beginning to crumble.

Very few of those participant in the current "turn toward the body" seem to notice the wider, more subversive implications of their work. While they assert that the entirety of the body is integral to the mind, surely (they assume) it is only the *human* body that has this privilege, and not the body of an elk, or an aspen grove, or the dense flesh of the ground itself. Surely the gushing body of a river, or the ebb and flow of the breeze, has no real part in intelligence!

It is here that their timidity contrasts with the far-seeing audacity of Spinoza. He alone saw that the human mind could never be reconciled with the human body unless intelligence was recognized as an attribute of nature in its entirety. To Spinoza, every sensible phenomenon had its own mental aspect; every tangible body within the material world was also an idea within the vast, encompassing intelligence that was known inwardly (to some) as God and outwardly (to all) as nature.

Despite his outmoded methodology, laden with geometrical terms (with numbered definitions, axioms, propositions, and corollaries), the heart of Spinoza's intuition remains prescient. For once we acknowledge that our awareness is inseparable—even, in some sense, indistinguishable—from our material physiology, can we really continue to maintain that mind remains alien to the rest of material nature? Consider how completely your sentient body is entangled in the crowd of creatures and elemental forces that enfolds you. Consider how thoroughly your organism is dependent upon these other lives: how your flesh is nourished and sustained by the plants whose leaves or fruits you ingest, by the other animals whose muscles you may eat, or whose milk you may drink, or whose carefully laid eggs firm the chocolate cake you nibbled on last night. And ponder, too, how your life sustains others in turn. Consider how your breath is taken up within the green chemistry of these grasses and whispering conifers, and how their exhalations add themselves to the swirling winds that embrace you, from which your lungs must drink, again and again, to fire your gestures and your streaming thoughts. Notice the pleasure that your fingertips find in certain wave-polished stones, the giddy thrill that your open palm, held out the window of a car, draws from the rush of wind that blasts against it. Notice the way your ears empty themselves toward the song of a wood thrush, or the manner in which your eyes are lured, like bees, by the interior azure of certain blossoms. The human body is not a closed or static object, but an open, unfinished entity utterly entwined with the soils, waters, and winds that move through it—a wild creature whose life is contingent upon the multiple other lives that surround it, and the shifting flows that surge through it.

Of course, our awareness is hugely informed by the exchange we carry on with those of our own species. Profoundly so. Yet the unique savvy of the human creature—the goofy grace of our gestures and, at times, of our thoughts—has also been shaped by our relation to the countless other earthly powers, both familiar and frightening, with whom we have coevolved. Throughout the obscure eons of our species' unfurling, it was these darkly dangerous and many-voiced surroundings that necessarily compelled our body's curiosity and desire, posing disturbing puzzles for our senses, prompting the human creature to pause, to ponder, and ultimately to reflect. And so it is misleading, today, for cognitive scientists to focus solely upon the human body in isolation from that larger matrix—as though our capacity for conscious reflection were somehow born only in interaction with ourselves, rather than with our world.

Today's intrepid researchers have yet to notice that the human body, in itself, is no more autonomous—and no more conscious—than an isolated brain. Sentience is not an attribute of a body in isolation; it emerges from the ongoing encounter between our flesh and the forest of rhythms in which it finds itself, born of the interplay and tension between the world's wild hunger and our own. The impulse toward thought grows from the gap between our thirst and an unexpectedly dry creekbed, as our curiosity finds consummation in the magnetism between our tongue and a prickly hedge studded with blackberries. Human awareness could not exist without a human body, true, but it could no more exist in the absence of ground, leaf, and flowing water. Mind arises, and dwells, between the body and the Earth, and hence is as much an attribute of this leafing world as of our own immodest species.

❦

My first two years at college left me ensnared in words, in snatches of text and snarls of argument that gripped my thoughts but smothered my ability to feel the world in ways less verbal. I walked out after the second year and spent the next wandering as a street magician through the cities of old Europe, imbibing the twang and rumble of other languages, letting those odd styles of speaking loosen my gestures and shake up my perceptions. How big the world was beginning to seem! Returning to college, I again felt pressed into a too tight pattern by the texts and taxonomies, and found myself yearning for a way back out of the words into the wildness of things. When classes ended that year, I walked onto a New England highway and stuck out my thumb, catching a series of rides westward across the continent. Once the snow-decked ridges of the Rocky Mountains lifted themselves from the horizon I stared in happy amazement and stepped out of the last truck, thanking the driver and making my way into a town where I began performing magic, for tips, in the local bars. After a fortnight I'd earned enough to purchase a decent tent and a sleeping bag; I shouldered my backpack and walked into the mountains.

I had camped a fair amount with friends and family when growing up, but had never before pitched a tent alone in the back-country. As my legs carried me past the last of the phone lines and into the thick of the forest, as the shadows deepened and the exclusively human world fell behind me, a great remembering shuddered through my muscles, as though a soul long buried were striding to the surface. My own real creaturely life, at last, was what was smelling those dank scents and hearing the pines rub against each other. Over the following days and nights, camping under high passes in snowfields agleam with moonlight, or hiking among rock-studded meadows articulated with gurgling rills, I found myself sliding through a vast array of feelings and moods, following thoughts as they meandered and fed into other insights and knowings—yet very few of these thoughts were embodied in words. I was thinking, yes, but in shifting shapes and rhythms and dimly colored vectors, thinking with my senses, feeling my way toward insights and understandings that had more the form of feelings blooming in my belly than of statements being spoken within my skull. A kind of spell had been broken; the school-hardened skein of words had softened, had loosened, had let me squeeze through and leave it behind like the grid of power lines I'd left at the edge of that wilderness, and so I was alone with my breath as the woods creaked around me and turquoise beetles climbed the grass blades and the owls hunkered down and waited for dusk.

It was there, in that solitude, that I first noticed how the drift of my thoughts was instilled and steadily carried by subtle alterations in the landscape. Walking in the woods kept my thoughts close and complexly patterned, while emerging into the wide meadows opened my ponderings out onto broad vistas of feeling, yielding insights into the expansive arc of my life and of the world's unfolding. As the new sun climbed above the peaks I could see its light roll toward me across the field, igniting the grasses and the scattered wildflowers, charging the air with warmth as it approached until it burst upon me as well, gleaming my naked surfaces, wrapping me in the grin of morning as it rolled on through the stony valley, and I could feel the petals of my brain slowly opening to meet that warmth. Sunlight was a mood that colored all my thoughts as I hiked, although if there were no clouds to break the heat I noticed my reflections melting together by midafternoon. A kind of languor then seeped into my muscles, blurring the keen edge of my thinking, as a hazy dream logic began to infect insights that'd been perfectly precise an hour earlier.

The ways of the mind seemed more manifold and mysterious here than I'd ever realized. I was beginning to glimpse a complex array of images for mind itself, visible patterns of mental process far more fitting than the neurological categories and mechanical descriptions I'd been inundated by in my psychology classes. Here, all around me, was a field of patterned metaphors as precise as one could want for the dynamic life of the psyche.

I had always been struck by the way that certain thoughts can arise unbidden, suddenly slipping into awareness in a manner that compels my attention; I now noticed how kindred such thoughts are to individual birds that glide into view, some flapping across the sky on the way to elsewhere, others spiraling and circling, trying to espy their prey, another swooping down to alight on a near branch and survey the scene. Still other winged thoughts are entirely hidden within the trees; although I hear their calls at regular intervals I hardly notice them anymore, and feel them only as a vague disturbance in the thickets.

Other notions move mostly beneath my conscious awareness, like subterranean streams that surface here and there among the clumped grasses, quiet undercurrents of thought that find their way into visibility now and then before slipping back into the rocky depths and freeing my focus to wander elsewhere. Still those unseen waters keep flowing, nourishing the soil underfoot.

Many taken-for-granted knowings are rooted and steadfast in the mind, like the firs and Engelmann spruces interspersed at the meadow's edge, or the aspens clustered in groves; I need the resilient strength of these powers to lean or climb on when structuring certain arguments or public assertions. Innumerable other insights, simple, small, and obvious, replenish themselves steadily like the uncountable blades of grass sprouting from the soil; among them lie the bric-a-brac of old and bedrock knowledges now broken and scattered, angling this way and that, feldspar and quartz glimmering on their surfaces.

There is a summer pulse to my awareness, too—an inward beat inseparable, now that I notice it, from the thrum of the crickets.

Mind, here in this high valley suspended beneath the blue, seems a vast thing, open and at ease. The thoughts that soar into view, the sedimented knowings, the bright blossoms of sensation are all held, here, within an encompassing equilibrium, permeated by a silence that swells and breathes with the cycles of light.

As the afternoon deepens and shadows lengthen among the trees, the sheer face of the peak at the southeast edge of this valley begins to glow with a ruddy light, and different strata of stone become apparent, the layers slanting smoothly up toward the southern sky until a point where the strata buckle and fold, down and up, a great fold in the long story of that massive cliff. Of what event is it the trace? I do not know whether that visible buckling of the layers bespeaks a sudden cataclysm or a long, slow application of pressure. I know only that the rock face speaks vividly to my senses, attesting to an uncanny depth of time—to a packed density of experience anciently undergone by matter itself, a density that now underlies and quietly supports our brief life in this open moment. Those folded layers speak, as well, of a startling malleability in even the most stable accretions of past experience. As though the solidified past of a continent, or the long-settled structure of a particular person, may still be upended and reshaped by huge and implacable forces that suddenly obtrude from outside any known frame of reference.

Just as (in a less powerful sense) a dark cloud is unexpectedly gliding into visibility from behind the high escarpment at the western perimeter of this bowl. Only moments ago the edge of that cloud seemed an innocent clump drifting on the verge of an unbroken blue; now that same cloud is rapidly swallowing what little is left of the open sky. Every mountain wanderer has learned this stealth lesson of the heights, the way that clouds can gather entirely hidden from view, secretly massing behind a high ridge that deceptively bounds a clear sky. Then a leaden shroud abruptly swings into view overhead, unleashing a stab of lightning upon the rocky precipices before breaking open and drenching your limbs as you lean over your unbuckled backpack, desperately digging in its depths to find the raincoat you idiotically stashed at the bottom.

This time, though, I don't take off my pack—I stand there inhaling the darkening shade, eager to welcome the rain, if it comes, and let it wash the sweat from my chest. The crickets are still thrumming, but more sporadically; the whole tenor of this place has changed now that we're hidden from the sun. The repeated cry of a jay, which I'd not noticed while it was sounding, now seems conspicuous by its absence. Throughout the meadow the boulders seem crouched in readiness, their lichen-scabbed surfaces sampling the air. In the midst of all this alertness something else is waking, is stirring and shifting, and then I realize it's the forest needles skirmishing as a breeze starts to flow down from the western ridge, pouring itself through the skein of branches and now across this open meadow, the grasses swirling in the gusts, purple lupines and columbines bouncing as the wind rises.

Isn't this how mind itself alters—one's pleasureful pondering abruptly transformed by the unexpected intrusion of a difficult memory, or by the pressure of a long-forgotten responsibility suddenly making itself known? The chemistry shifts, the mood changes; every new thought is now shaded by the gray blanket overhead, or else enlivened by the prospect of rain.

If that dark ceiling does crack open, then this field will soon be a sodden marsh impossible to walk through. I survey my options, then tighten my pack and make my way between rocks and wildflowers toward the woods. The ground's a bit higher there, and more solid, buttressed by roots a hundred times thicker than those twined in mats beneath the meadow. When I arrive among the trees I swing the pack from my shoulders, balancing it a moment on one knee as I extract my arm, then lower it to the ground, leaning it against a tall pine. I peel off my sweat-soaked T-shirt and hang it from a corner of the pack, then lean myself against the other side of the same trunk, the bark scraping against my spine. Overhead the treetops are tossing in the wind; down here all is weirdly peaceful and unperturbed, another world. After a stretch of breaths, I sense the thick trunk of my lean-tree subtly swaying my torso from side to side. Perhaps it is only my imagination? No, now I can feel it quite clearly. I gaze the neighboring trunks at eye level—subalpine fir, Engelmann spruce—trying to discern if they, too, are moving. If they are, I cannot tell.

The tree continues to rock me, gently, rooting my spine in the shallow soil and letting thoughts rise, fall, and dissipate. I wake to the strange sensation of being watched. But how, by whom? Then

I glimpse it: just behind the near trees, a small movement. Like that of a leaf, flapping, except not quite—and then I recognize that, yes, it is an ear, flicking. A rather large ear. And just below it, a round eye is looking back at me—although perhaps not seeing me, for now a single antler rises into view as the buck turns its head down to continue browsing. I lean right in hopes of a clearer view: the deer looks up again. After a long moment, he slowly lowers his head, then suddenly raises it again, as if to catch me. I am per-fect-ly still, my breath less than it would take to stir a spiderweb. Finally the buck seems convinced; he returns to browsing, shifting his stance to munch another patch of leaves. Slowly, now, the deer moves toward my left, gradually emerging from behind those trunks, yet still masked by a spray of branches. Only then do my eyes glean an outline of its body, and receive something of that taut grace—the brown flank, the neck extending itself forward, then resting, and then gliding forward again, as one leg after another steps through the air. Not twenty feet in front of me, its tawny body slips easily among the boughs. I twist my torso to gaze it more directly, but . . . the buck has vanished. I stare dumbfounded through the lattice of trunks, tilting my body this way and that, listening for the thumping of hooves—yet hear only the surge in the high branches, and the utter quietude in front. I step away from my tree, striding forward several yards as I push past various limbs, yet nothing shifts or sounds in front of me. Off to my right, a faint chirping in the branches. But the deer is gone.

Impossible. Wasn't a rather large, warm-blooded creature stepping between these very trees just a moment ago? How could it simply disappear, as though it were a mere figment?! Then again, I suppose I never did see it clearly—was watching its movements rather than staring at its stable presence—but still . . . it *was* present, was it not? Ahh, look at this leaf, and this other one, both clearly nibbled on, and if I look down—of course: here are the tracks, the double teardrops neatly pressed into the dirt. And look closer: each rear hoof, slightly smaller, lands in the same print that the front hoof just left—the clear trace of that silent gait.

❧

But it was odd how familiar that fleeting encounter felt to me on that day. It seemed the precise image, once again, of a certain style of thought, a living metaphor for the way the mind moves. How many times has it happened, in the course of going about my work, that a potent thought will make itself felt at the edge of my awareness—an insight whose rightness I can sense even before I know its content—yet when I turn to fix my attention upon it, I find that it has already vanished, leaving only a vague flavor or trace? It's the sort of thing one easily forgets. Yet it has happened frequently enough that now I'm alert to it. I often feel such an insight foraging in the thickets of my consciousness, and can sense its elegance, somehow, its sleek profundity and pertinence nibbling at the very periphery of my attention. Yet when I try to focus my gaze upon it, it slips back into the shadowed depths, leaving me with only a curious whiff of feeling—the spoor of a ghost—unable to say what that insight was, or even to what it pertained.

The frustration entailed by such phantom thoughts had disrupted my youthful confidence in my own ability, as a philosopher, to reflect upon matters with some accuracy. But I was now beginning to realize another possibility: that ideas had their own lives independent of mine. That indeed some vital ideas were like creatures wholly unaccustomed to human contact, wild notions whose robust elegance and vigor required that they keep their distance from those who might strive to define or domesticate them, twisting them from their native habitat. The real space of the mind no longer seemed like an indoor room—like a laboratory, or an operating theater, wherein subjects could be immobilized and exposed to the cool light of analysis—but rather more like a wooded valley, or a meadow shadowed by cloud: an open terrain through which I was wandering.

Deer, of course, are quintessential creatures to human culture. For much of civilized humankind, deer are the most familiar animals of the wild. They dwell just beyond the edge of our streetlights, unfazed by our machinations; they slip through our settlements when we're at rest, and raid our gardens. The name "deer" (in its original spelling, "deor") was once the Old English word for "animal" in general (as the term "wilderness," or "wild-deor-ness," once meant simply "the place of wild animals"). My dawning suspicion that certain ideas were like deer, visiting our awareness in much the same way that wild deer make contact with us—graceful, shy, lingering at the edge of our awareness, yet slipping back into the forest if too willfully focused upon—suggested that other ideas intersect our awareness more in the way of *other* animals. Some of them slither mostly unseen through the grass, sporting different hues and patterns in keeping with their favored haunts; others bask on warm rocks in the midafternoon, only to skitter away at our incautious approach. Certain lightweight thoughts flutter in the air around us, so small and erratic we easily neglect to notice them, while other more muscled notions lope unexpected across our roads, marking their passage with scent or scat. Dusky intuitions hunch on the middle branches, their eyes closed, listening intently for the soft rustlings of those other, tinier hints upon which they feed. Some of the most versatile ideas are entirely multiple, embodied by a thousand humming lives lofting and veering in concert; others are solitary powers, reticent and sly, able to keep their own counsel. There are insights we come upon only at the edge of the sea, and others we glimpse only in the craggy heights. Some prickly notions are endemic to deserts, while other thoughts, too slippery to grasp, are met mostly in swamps. Many nomad thoughts migrate between different realms, shifting their habits to fit the terrain, orienting themselves by the wind and the stars.

❦

This manner of loosening conventional conceptions of mind, playful as it seems, gains new sustenance and substance when one learns that the word "idea" and the word "species" were once synonymous terms—one Greek, the other Latin—for the same enigmatic phenomenon! Both "idea" and "species" derive their current meaning from the ancient use of a single Greek term, "eidos"—a word that originally signified "the visible look, or outward form" of a thing. It was this word, "eidos," that the Greek philosopher Plato employed to speak of the immaterial and unchanging "forms" that alone, according to him, truly exist—the eternal "ideas" that can be known only by the pure, rarefied mind or intellect. (In ancient Greek, "idea" was simply the feminine version of "eidos.")

In the philosophy propounded in Plato's later dialogues, the palpable trees or clouds or toads that we perceive with our bodily senses are but ephemeral manifestations of universal, bodiless archetypes, the Platonic ideas—the ideal tree, the ideal cloud, the ideal toad. According to the Socrates of Plato's dialogues, whenever we ponder an individual oak tree, that which is genuinely knowable of that oak is not the sensible tree but the essential form, the idea, the *ideal* oak tree of which each individual oak reminds us. (The word "ideal" is simply the adjectival form of "idea.") The palpable tree is subject to change, to growth and decay, and hence has only a very provisional and insubstantial existence. The ideal oak, untainted by materiality, is eternal and unchanging, and hence—according to Plato—it alone can truly be said to exist. As we come to recognize these transcendent essences, or ideas, behind their material instantiations, and as we learn to contemplate these universal ideas in their purity, our mind gradually frees itself from the prison of the body and attains that transcendent, eternal state that is its truest home.

It was thus via Plato's philosophy that the word "idea" spread into European languages, as the primary term for the immaterial stuff that the mind ponders, slowly evolving the various nuances of meaning that the term has for us today.

Meanwhile, Plato's most precocious student, the protobiologist Aristotle, began to deploy his teacher's term in a subtly different manner. Aristotle challenged Plato's claim that the universal "ideas" (or forms) have any separate existence apart from the particular beings that materially embody those ideas. He questioned, in other words, the Platonic teaching that the forms "oak" or "toad" have their own utterly transcendent and prior existence independent of particular toads or particular oak trees. For Aristotle, the universal form of oak tree was simply that quality that all oaks had in common with one another; it was that which made them oak trees, rather than maples or pines. When we carefully contemplate a particular tree, according to Aristotle, it is this universal form that is received or recognized by the rational intellect. But this universal has no transcendent existence apart from the individual trees that exemplify it. Hence Aristotle began to employ his teacher's word, "idea," to speak of the group of individuals that share a common form, those individuals that constitute a particular kind of tree, a particular kind of plant or animal. And it is Aristotle's usage of Plato's term "idea" that was precisely translated into Latin by the word "species"—a term that originally signified, like the Greek "idea," the outward form or "look" of any entity.

In Aristotelian usage, then, first the Greek term "idea" and then the Latin term "species" came to mean a collective group of individuals that share a common form. Through the subsequent development of European languages and the emergence of modern English, the word "species" has retained much of its earthly, Aristotelian meaning, while the word "idea" reverted to its more ephemeral Platonic meaning, as an immaterial image or thought pondered by the mind.

This quick glance at the early history of these two terms reveals that the diverse flapping, leaping, and leafing forms that we now call "species" were once experienced (by some of our most canny ancestors) as presences entirely akin to "ideas." This, of course, because "ideas" were then held to have a much more independent and impersonal reality than they do in our time. Today "ideas" appear to have lost much of their universal, transcendent character—they no longer seem to be realities existing entirely independent of us. Like thoughts, notions, and insights, "ideas" are now assumed to be the mostly private ephemera of an individual mind, inhabitants of an interior zone of reflection that is unique to each person. You have your ideas, and I have mine. Sure, there may be some ideas that we hold in common, but that is because they have been taught to us both, or because we have agreed upon them—not because they subsist out there independent of human culture and its conventions. "Species," meanwhile, has lost all apparent association with the intellect. In contrast to the internal character of my thoughts and ideas, species seem entirely objective aspects of the external material world.

Yet how powerfully those rooted and antlered forms echo the manifold shapes that still move within our mind! It is this resonance that was communicating itself to me in the course of my solo walkabout in the Rockies, gradually confounding the clear distinction between inner and outer worlds, as the elfin mood of dusk blurred the boundaries between daytime and night, dissolving the sharp edges of trunks and softening the boulders.

With the first hint of dusk in those high valleys, various flying insects launched into the air, including large flies that would bedevil me where I sat cross-legged, usually on a boulder, watching the light fade from the upper ridges. It was during those crepuscular hours that I discovered a useful technique for disbursing such unnerving, whirring beings when they disrupted my contemplations. (It's a technique I still use to this day, although only when I'm off in the backcountry, since it requires a focused clarity and precision tough to muster in the middle of town.) Whenever a fly's feverish orbit around my head begins to unravel my poise, tearing apart with its buzz the simple calm of my mountain ponderings, I search my inward awareness for any small but annoying thought whose unpleasantness matches, in some way, that monotonous sprite buzzing again and again past my undefended ears. Oddly, such an irksome thought always seems easy to find at such a moment—as though the fly's buzzing had already drawn forth that nagging notion from the depths of my awareness. Once I've become conscious of that thought, I shift my attention to the air just in front of my face, watching that near space through which the fly weaves, again and again, trying to catch its trajectory in my gaze, waiting patiently for the serendipitous moment when I will manage, for an instant, to focus my eyes upon the fly. As soon as I do so, in that split second of visual contact, I offer my vexing thought to the insect. At which point the fly—to my considerable amazement—buzzes away! Sometimes my eyes have been able to follow the fly as it departs into the distance before me, becoming smaller and smaller until it dissolves. I am left sitting on the rock, unburdened of both bug and bothersome thought, released into the ringing silence.

I suppose it was only by leaving behind, for those weeks, the compulsion to communicate with words that I became so vulnerable to the expressive power of all those other-than-human styles of sensitivity and sentience. It was not merely the polymorphous play of rhythms, the syncopation of shapes that swerve and sprout beyond the confines of the city, but also how startlingly immense the land became when I encountered it without the steady filter of words, the discovery that I was palpably immersed in a field of unfoldings so much wider than myself and my intentions. It was not just the resonant metaphors offered by stones and grasses and muscled creatures, but also the rightness, somehow, of recognizing mind as a broad landscape within which I was wandering, a deep field with its near aspects and its distances, its moods shifting like the weather. For surely mind has its depths: memories buried, for instance, beneath the ground of our current awareness, or recent insights momentarily concealed behind the close matters we're obsessively stuck on.

There seemed something *more than metaphoric* here, something strangely right about this resonance between thought and the earthly terrain. For clearly there's something about the psyche that exceeds us and overflows all our knowings, confounding every notion of mind as a self-contained space within our head. Certainly, I still felt that there was an interior quality to the mind. But my encounters with other styles of sentience were loosening the conception of my own mind as a closed zone of reflection, stirring long-slumbering memories from my earliest years of life, bringing faint whiffs of a forgotten intimacy between awareness and the elemental earth. As though the leap and vanish of a deer into the forest or these other movements of shadows and grass and rain were not merely metaphors but part of the very *constitution* of the mind, of its real structure and architecture. And it was then that a simple thought burst upon my awareness—not like a bolt of lightning, but rather like a gentle rain beginning to fall around me, soaking my head and my chest, moistening the ground and raising its mingled scents to my nostrils: What if mind is not ours, but is Earth's? What if mind, rightly understood, is not a special property of humankind, but is rather a property of the Earth itself—a power in which we are carnally immersed?

What if there is, yes, a quality of inwardness to the mind, not because the mind is located inside us (inside our body or brain), but because we are situated, bodily, *inside it*—because our lives and our thoughts unfold in the depths of a mind that is not really ours, but is rather the Earth's? What if like the hunkered owl, and the spruce bending above it, and the beetle staggering from needle to needle on that branch, we all partake of the wide intelligence of this world—because we're materially participant, with our actions and our passions, in the broad psyche of this sphere?

It was a thought that sparked and rippled along the whole surface of my skin, as though all the pores of that smooth membrane were opening at once, as though the skin itself were coming awake. Or as though I myself was awakening as that shimmering membrane at the outer bound of my body—as this luminous shape resonating with the other shapes that surround me: rocks, grasses, trees—each contour a sparkling surface of metamorphosis and exchange between that being and the charged air around it, as a sort of eros played between us, an electric tension binding us all into the expansive Body of the place. The sensations along my skin were subtly shifting, I realized, in tandem with the changing scents of soil and wood and dusk as these wafted in at my nostrils, my own tenor transforming with each alteration in the breeze, my face awakening to the innumerable nuances drifting in the vorticed air around me. And my eyes widened with the recognition that a similar interchange was going on throughout the whole invisible ocean of air as it riffled through the forest needles and twisted the grasses and washed against the rock escarpments, each gust carrying chemical insights between the insects and the swaying blooms of this valley, every breeze exchanging pollen between the trees. I was alive in a living field of experience.

The full-bodied alertness I now felt seemed no more mine than it was the valley's; it seemed a capacity of the land itself that had been imparted to my body. Much as the life then beating in my chest and rolling through my veins was fed by the mountain air of that place (by the calm or blustering atmosphere that surrounds us wherever we are, and that we've been imbibing, ceaselessly, since the moment of our birth). Is not awareness, too, a kind of medium or atmosphere—a capacity that blooms within us, swelling and subsiding, only because we are penetrated by it, encompassed by it, permeated? Are we not born into mind as into an unseen layer of the Earth, gradually opening ourselves to the nourishment of this medium, adapting ourselves to its lunar rhythms, aligning ourselves with the way it glimmers and sings in our particular species?

Mind as a medium, yes. As an invisible and ceaselessly transforming layer of this world, a fluid medium that permeates our bodies. As though the air itself were aware! As though the unseen air that enfolds us, and circulates through us, were the very stuff of awareness. Yet not merely the air considered as a mix of gases; for isn't there also a kind of sparkle and hum to awareness, an electric quality—as though it were charged with a kind of tension, or snap? Not the element of air alone, then, but the element of fire as well, mercurial—a subtle flame dispersed throughout the medium. Much like the way the sun's radiant fire is buffered and broken up by Earth's atmosphere, dispersed and scattered among countless molecules of gas, and thus muted, transmuted, into a medium suitable for spruces and bears and belted kingfishers. Not exactly the air of this world, then, but the way sunlight lives in this air—this unseen medium charged with fire. This awakened atmosphere. This awareness in which we and the mountain trees are immersed. This mind.

Interesting poetry, perhaps. But why should we take such a curious claim regarding the nature of mind as anything other than poetry—anything other than a momentary set of metaphors used to convey a purely personal experience? If such an assertion purports to carry more than a merely personal meaning, then surely the methodology of science must be brought to bear, testing it as a hypothesis subject to proof or disproof. Clearly, with regard to the stuff of consciousness, science itself must set the rules!

Unless such recourse to science, as the sole arbiter of collective truth, is thoroughly misplaced when we come to matters of mind or consciousness. For consciousness is a quintessentially quicksilver phenomenon, impossible to isolate and pin down. As soon as we try to ponder the character of consciousness, we discover that it's already escaped us—for it is really the pondering that we're after, rather than the thing pondered. We find ourselves unable to get any distance from awareness, in order to examine it objectively, for wherever we step it is already there. Perhaps we can at least say clearly what awareness is not, and can thus narrow down an assessment of its true character. Yet here, too, we are foiled, for any entity that we notice or name—in hopes of separating it from the substance of awareness—is, alas, already entangled in that substance.

Awareness, or mind, is in this sense very much like a medium in which we're situated, and from which we are simply unable to extricate ourselves without ceasing to exist. Everything we know or sense of ourselves is conditioned by this atmosphere. We are intimately acquainted with its character, ceaselessly transformed by its influence upon us and within us. And yet we're unable to characterize this medium from outside. We are composed of this curious element, permeated by it, and hence can take no distance from it. For this reason alone, awareness will never be fully amenable to the methods of empirical, objective science. While it conditions and makes possible every scientific endeavor, awareness will never entirely be made over into an object for science. It remains an enigma, a wonder, not by virtue of any miraculous quality, but by the plain and obvious fact that we are in it, and of it.

By suggesting that we all inhabit a common awareness, I am not claiming that your thoughts and dreams are the same as my thoughts and dreams, or that my awareness is the same as that of a magpie, or a coyote, or a mountain hemlock. For we each engage the wider intelligence from our own angle and place within it, each of us entwined with the breathing Earth through our particular skin. This ecological articulation of mind thus yields *a radical and irreducible pluralism.* It is only as palpable bodies that we participate in the immersive mind of this planet, and as your body is different from mine in so many ways, and as our limbs and senses are so curiously different from those of the pileated woodpecker or the praying mantis, just so are our insights and desires richly different from one another. Each creature's experience is unique, to be sure, yet this is not at all because an autonomous awareness is held inside its particular body or brain. Sentience is born of the ongoing encounter, the contact, the tension and entwinement between each body and the breathing world that surrounds it. While each of us encounters only a corner of this world, it is nonetheless the same outrageous world—the same Earth—onto which our various senses open, the same inscrutable Earth that each engages with its fingers or its feathered wings, with its coiled antennae or its spreading roots.

Now, if my experience arises from the interaction between this body and the elemental field that surrounds—if my awareness is thus provoked by the same vast and compassing planet that calls forth and sustains your awareness—then how much sense does it make to continue to say that each of us has our own self-sufficient mind neatly tucked inside our particular head, while the surrounding Earth is just an object, utterly bereft of all intelligence? Is it not far more parsimonious to admit that mind is as much a property of the windswept Earth as it is of the myriad bodies that flock and congregate upon its ground? Shall we prolong the painful split between mind and body by continuing to neglect our carnal entanglement with this immense Presence, or shall we finally heal that age-old wound by acknowledging Earth's implicit involvement in all our experience—as the solid ground that supports all our certainties, and the distant horizon that provokes all our dreams?

As long as we overlook our ongoing intercourse with the Earth, then we are compelled to double each human body with a separate mind—since mind's wild fluidity, its creativity and openness, seems then of an entirely different order than our corporeal, physical frame. But as soon as we recognize that our bodies are always intertwined with the broad flesh of the Earth, and that our conscious experience is sustained and steadily informed by that very involvement, then the need for a multitude of individual, immaterial minds drops away . . . There is a profusion of individual bodies; there is the enveloping sphere of the planet; and there is the ongoing, open relation between these. The fluid field of experience that we call "mind" is simply the place of this open, improvisational relationship—experienced separately by each individual body, experienced all at once by the animate Earth itself.

I do not mean here to minimize the effect that interpersonal relationships have upon our human experience of mind, or to neglect the patterning influence that any culture, and language, has upon our awareness. I mean only to point out a truth all too easily forgotten in this technological era, which is that culture can impose its patterns only within the constraints set by the biosphere itself. Family and friends have their pull upon us, as do our various traditions of behavior and belief, yet none of these can dissolve the inexorable influence that the planet has upon our experience. The languages we speak do indeed limit our conscious conceptions of the world, yet no words can dispel the power that gravity and rain and the steady need for Earth's air have upon the form of our bodies, the shape and texture of our thoughts.[n]

We may, of course, continue to speak of mind as an excellence utterly unique to our species, a capacity that springs us free from our embedment in the earthly community of animate forms. We may continue to hold that the rustling of experience—of exultation and grief, of compassion and confusion—is a purely *human* thing, and hence that the felt stirrings undergone by other creatures are mute and expressionless phantoms, automatic reflexes entirely closed to creative nuance. We may persist in the modern assumption that a dammed-up river has no impulse of its own, that the mystery of mind has no pertinence to a meadow or a dying woodland, or to a nascent thunderstorm gathering above the plains, seething and crackling with energy before the first bolts snap the air.

But by doing so, we seal ourselves into a numbing solitude—a loneliness already settling around us as the complex creativity of forests gives way to the numbered productivity of even-aged tree farms, as the diverse riffs of songbirds steadily fade from the soundscape, and the wild, syncopated chant of the frog chorus that once rocked the fields every spring dwindles down to the monotonous hum of a single street lamp. Do we really believe that the human imagination can sustain itself without being startled by

[n] From this perspective, we could say that the brain itself is an introjection of the earth, an analogue or avatar of the planet happily riding atop our spine. (An introjection is the opposite of a projection—something outside oneself is replicated within.) Not only does the sensorial environment (with its elemental mix of animals, plants, and landforms, its available foods and its lurking predators) steadily exert a selective pressure upon the patterned associations between neurons—but also the large-scale structures of the brain (the cerebal tissues and organs found across persons and even across species) have formed themselves in response to the most stable structures of the perceptual field, to the openess of the horizon and the density of the ground on a planet with this specific gravity, to the chill nourishment of rain and the steady singing of birds . . . The brain is an introjected earth.

other shapes of sentience—by redwoods and gleaming orchids and the eerie glissando cries of humpback whales? Do we really trust that the human mind can maintain its coherence in an exclusively human-made world?

Perhaps the broad sphere, itself, needed our forgetfulness. Perhaps some new power was waiting to be born on the planet, and our species was called upon to incubate this power in the dark cocoon of our solitude. Our senses dulled, our attention lost to the world, we created, in our inward turning, a quiet cave wherein a new layer of Earth could first shape itself and come to life. But surely it's time now to hatch this new stratum, to waken our senses from their screen-dazzled swoon, and so to offer this power back to the more-than-human terrain. The cascading extinctions of other species make evident that the time is long past ripe. The abrupt loss of rain forests and coral reefs, the choking of wetlands, the poisons leaching into the soils, and the toxins spreading in our muscles compel us to awaken from our long oblivion, to cough up this difficult magic that's been growing within us, swelling us with pride even as the land disintegrates all around us. Surely we've cut ourselves off for long enough—time, now, to open our minds outward, returning to the biosphere that wide intelligence we'd thought was ours alone.

Mind, then, as the steady dreaming of this Eairth, the unseen depth from whence all beings draw their sustenance. Mind as wind, this whooshing force, the medium that moves between and binds all earthly beings. This mystery rolling in and out of breathing bodies, invisible: wafting up from the soil and gusting down from the passes in vortices and curling waves that splash against the grasses. A fluid receptacle, clear as crystal, for the inpouring fire of sunlight; the pellucid nighttime lens through which our eyes receive the stars.

Sentience was never our private possession. We live immersed in intelligence, enveloped and informed by a creativity we cannot fathom.

MOOD

(Depth Ecology III)

After all, anybody is as their land and air is. Anybody is as the sky is low or high, the air heavy or clear and anybody is as there is wind or no wind there. It is that which makes them and the arts they make and the work they do and the way they eat and the way they drink and the way they learn and everything.

—GERTRUDE STEIN

As soon as we breathe out, letting mind flow back into the field that surrounds us, we feel a new looseness and freedom. The other animals, the plants, the cliffs, and the tides are now participant in the unfolding of events, and so it no longer falls upon us, alone, to make things happen as we choose. Since we are not the sole bearers of consciousness, we are no longer on top of things, with the crippling responsibility that that entails. We're now accomplices in a vast and steadily unfolding mystery, and our actions have resonance only to the extent that they are awake to the other agencies around us, attuned and responsive to the upwelling creativity in the land itself.

When we allow that mind is a luminous quality of the earth, we swiftly notice this consequence: each region—each topography, each uniquely patterned ecosystem—has its own *particular* awareness, its unique *style* of intelligence. Certainly the atmosphere, the translucent medium of exchange between the breathing bodies of any locale, is subtly different in each terrain. The air of the coastal northwest of North America, infused with salt spray and the tang of spruce, cedar, and fir needles, tastes and feels different from the air shimmering in the heat of the southwest desert. Each atmosphere imparts its vibrance to those who partake of it, and hence the black-gleamed ravens who carve loops through the desert sky speak a different dialect of squawks and guttural cries than the cedar-perched ravens of the Pacific Northwest, whose vocal arguments are often instilled with liquid tones. Likewise, the atmosphere that rolls over the Great Plains, gathering now and then into swirling tornadoes, contrasts vividly with the blustering winds that pour through the Rocky Mountain passes, and still more with the mists that advance and recede along the California coast. The specific geology of a place yields a soil rich in particular minerals, and the rains and rivers that feed those soils invite a unique blend of grasses, shrubs, and trees to take root there. These, in turn, beckon particular animals to browse their leaves, or to eat their fruits and distribute their seeds, to pollinate their blossoms or to find shelter among their roots, and thus a complexly intertwined community begins to emerge, bustling and humming within itself. Every such community percolates a different chemistry into the air that animates it, joining whiffs and subtle pheromones to the drumming of woodpeckers and the crisscrossing hues of stone and leaf and feather that echo back and forth through that terrain, while the way these elements blend is affected by the noon heat that beats down in some regions, or the frigid cold that hardens the ground in others.

Each place has its rhythms of change and metamorphosis, its

specific style of expanding and contracting in response to the turning seasons, and this, too, shapes—and is shaped by—the sentience of that land. Whether we speak of a broad mountain range or of a small valley within that range, at each scale there is a unique intelligence circulating among the various constituents of the place—a style evident in the way events unfold in that ecosystem, how the slow spread of a mountain's shadow alters the insect swarms above a cool stream, or the way a forested slope rejuvenates itself after a fire. For the precise amalgam of elements that structures each region exists nowhere else. Each place, that is to say, is a unique *state* of mind, and the many powers that constitute and dwell within that locale—the spiders and the tree frogs no less than the humans—all participate in, and partake of, the particular mind of the place.

Of course, I can hardly be instilled by this intelligence if I only touch down, briefly, on my way to elsewhere. Only by living for many moons in one region, my peripheral senses tracking seasonal changes in the local plants while the scents of the soil steadily seep in through my pores—only over time can the intelligence of a place lay claim upon my person. Slowly, as the seasonal round repeats itself again and again, the lilt and melody of the local songbirds becomes an expectation within my ears, and so the mind I've carried within me settles into the wider mind that enfolds me. Changes in the terrain begin to release and mirror my own, internal changes. The slow metamorphosis of colors within the landscape; the way mice migrate into the walls of my home as the climate grows colder; oak buds bursting and unfurling their leaves to join a gazillion other leaves in agile, wind-tossed exuberance before they tumble, spent, to the ground; the way a cat-faced spider weaves her spiraling web in front of the porchlight every summer—each such patterned event, quietly observed, releases analogous metamorphoses within myself. Without such tunement and triggering by the earthly surroundings, my emotional body is stymied, befuddled—forced to spiral through its necessary transformations without any guidance from the larger Body of the place (and hence entirely out of phase with my neighbors, human and non-human). *Sensory perception, here, is the silken web that binds our separate nervous systems into the encompassing ecosystem.*

Human communities, too, are informed by the specific sentience of the lands that they inhabit. There is a unique temperament to the bustling commerce and culture of any old-enough city, a mental climate that we instantly recognize upon returning after several years, and that we mistakenly ascribe solely to the human inhabitants of the metropolis. It is a result, we surmise, of the particular trades that the city is known for, or the dynamic mix of ethnicities that interweave there, or the heavy-handed smugness of the local police force. Such social dynamics, however, are steadily fed by the elemental energies of the realm—by the heavy overcast that cloaks the sky for weeks at a time, or the splashing speech of the river that rolls through downtown, tossing glints of sunlight into the eyes of all who walk near, or by the way that the greasy exhaust from forty thousand commuting cars interacts with the humidity of the summer air. The dismal social ills endemic to certain cities have often been stoked by the foolishness of urban designers who overlooked the particular wildness of the place, ignoring the *genius loci,* the unique intelligence of the land now squelched and stifled by local industries. A calloused coldness, or meanness, results when our animal senses are cut off for too long from the animate earth, when our ears—inundated by the whooping blare of car alarms and the muted thunder of subways—no longer encounter the resonant silence, as our eyes forget the irregular wildness of things green and growing behind the rectilinear daze.

Still, even the stone structures of a metropolis may become expressions of the *genius loci.* Old buildings regularly worked over by the sun, rain, and wind finally become gestures of the local earth. The very architecture of any city old enough to have negotiated with gravity, century after century, for the stance of its walls and the solidity of its foundations, is now a conduit for the pulse and power that rises steadily from the ground. The people who take up residence and work in that city are channeled into patterns of life appropriate to the realm by the edifices that surround them, by the worn-smooth cobbles of the winding streets and the slightly sagging structures within which they dwell.

Thus the old city of Prague, which straddles the Vltava River with its statue-laden bridges and its battlements, its narrow lanes and its Gothic towers, holds certain chthonic qualities in place even as the recently released tides of capitalism sweep across it. The strange and brooding psyche of the local earth is preserved and curiously protected in the hand-built substance of this town, where High Gothic structures from the fourteenth century jive with ornate Baroque edifices from the seventeenth, where somber Romanesque churches converse in muted tones with audacious Art Nouveau neighbors. Despite the corrosive power of the capitalist economy (usually a kind of universal solvent), an earthly magic dwells and moves within the city of Prague. It's a power that cannot easily be eradicated here, where the pagan Green Man still peers down from the shadowed corners of medieval churches (stone vines spiraling from his gaping mouth), where zodiacal clocks in the looming clock towers tick off the mythical hours, and ravens squawk in the branches above a cemetery so thick with tilting headstones that even the ghosts cannot squeeze between them. Given solid expression in the ornate structure and statuary of so many buildings, the telluric intelligence of the local earth had only to recede into the old stonework in order to survive the sterile era of Soviet communism, with its anti-aesthetic of mass efficiency, or to weather the new occupation by the culture of total commerce. From within those aged walls, the energies pulsing up from the ground continue to exert their influence upon all who live and work there. Hence the kabbalistic and magical arts still flourish in the city of Prague. I am acquainted with a molecular biologist there, respected for his research throughout Europe, who at night composes and interprets astrological charts for the country's intelligentsia. I have been befriended by an accomplished and erudite geologist in that city who regularly turns his attention to the arcana of geomancy in the off hours. Come evening, the live sounds of chamber music—Mozart and Smetana and Janáček—

echo through the narrow streets, while in the morning crowds of hand-carved marionettes dangle behind shop windows throughout the old town, their painted eyes peering at each passerby, awaiting the curious person who will finally bring their limbs to life. In old Prague, as in other venerable cities whose cobbles have been bloodied and buffeted by succeeding ages, the built scape of the city does not stifle the sentience of the land, but rather ensures that that power continues to vitalize the human community.

The intelligence that moves within the Czech landscape is very different from that which animates the city of Paris, as the wide sky reflected in the meandering Seine contrasts vividly with the low overcast often mirrored in the Vltava River. The disparity between the broad resplendence of the Parisian esplanades or the tastes of Parisian cuisine and the narrow alleys or flavors common to Prague expresses a difference in their human histories, to be sure, but also in the very ambiance of their respective terrains, and the specific fertility of their soils. Meanwhile, the contrasting bustle of Rome—the grand bluster of its mythic statuary and the convivial ethos of its many fountains—bespeaks the very different emotional climate of that peninsula, and of the elemental forces (the specific goddesses and gods) that compose the terrestrial psyche of that land.

For each land has its own psyche, its own style of sentience, and hence to travel from Rome to Paris, or from Barcelona to Berlin, is to voyage from one state of mind to another, very different, state of mind. Even to journey by train from Manhattan to Boston, or simply to walk from one New England town to another, is to transform one's state of awareness. Traveling on foot makes these variations most evident, as the topography gradually alters, mountains giving way to foothills, and foothills becoming plains, as the accents of the local shopkeepers transform in tandem with the shifting terrain. The texture of the air changes as the moisture-laden atmosphere of the highlands, instilled with the breath of cool, granitic caves and the exhalations of roots and matted needles, opens onto the dry wind whirling across the flatlands, blending the scents of upturned soil with hints of exhaust from the highway, and—especially strong in some places—the acrid smell of processed fertilizer.

Such alterations in the unseen spirit of the land are mostly hidden to those who make the journey by car, since then all the senses other than sight are held apart from the sensuous earth, isolated within a capsule hurtling along the highway too fast for even the eyes to register most changes in the disposition of the visible. Still, subtle clues drift into the cabin, now and then—the insistent stench of those fertilized fields, or the reek from an unfortunate skunk finding its way even into nostrils well insulated by air-conditioning. If we turn on the radio, then our ears, too, may be assailed by the shifting psyche of the land—the percussive hip-hop and blues of the city opening onto the lilting voices and plucked strings of country music (laced with funk in regions closer to Gotham and more plaintive as we roll into more rural spaces). Along broad stretches of the interstate, the wavelengths give way to a saturated array of Christian stations, with smooth- or gravel-voiced preachers citing chapter and verse with a rhythmic passion exceeding even that of the urban rap artists. This, too, is a register of the mind of that locale.

Certainly there are roads more conducive to sensory alertness than the highway, whose rectilinear, manicured margins induce a kind of stupor in the speeding driver, a steady trance of abstraction—an onrushing flood of memories and future concerns with few ties to the sensuous present. Unpaved, country back roads invite a more open and improvisational style of thought, one that stays bound, albeit loosely, to the twisting topography. This is a more meandering cogitation, vaguely punctuated by leaning mailboxes and rusting maroon tractors, by barking dogs racing the car or a kestrel motionless on a fence post.

Yet how much more thoroughly this land would feed our thoughts if we were not driving but rather strolling on foot along these lanes—or even pedaling a decent bicycle, the gusting wind swelling our lungs as our muscles work themselves against the slope!

If the automobile isolates our speeding senses from the land, the airplanes in which we fly abstract us entirely from the grounding earth. After checking our bags at the airport, we tighten our buckles and loudly levitate up out of the ecosystem, shaking our senses free from the web of relationships that compose the specific sentience of that place. Only to plunk down some hours later in an entirely different ecology—an entirely different state of mind—without experiencing any of the transitional terrain between them, without our nervous system being tuned and tutored for this change by the gradual changes in the topography as we move across it. It's a recipe for madness, don't you think?—the way we now force our body to slip from one mode of awareness into another weirdly different awareness, without undergoing the slow, perfectly calibrated transition between them (the gentle metamorphosis that the earth instills in our body as it alters around us). No wonder that we feel dazed and discombobulated after an extended flight! "Jet lag" we call it, dutifully resetting our watches, as though it were merely a consequence of entering a new time zone. As though it had nothing to do with abruptly finding ourselves in a world whose background colors, shapes, and smells diverge drastically from those where we were a few hours earlier. A new time zone? Well, yes, if by this we mean a place whose rhythmic *timing*, or pulse, is oddly other than that where we just came from—a zone whose specific dynamism tempts our skin and jangles our ears in weird new ways, crashing our sensory organization, forcing our nervous system to reorder itself as best it can. The sudden strangeness is jarring to our animal body, and especially rattling when we're compelled to adapt to the new circumstance in a matter of minutes.

Hence, after flying in jets for several years, many adapt by simply blunting their senses, numbing themselves to all but the most homogeneous facets of a place, taking refuge in only the most fabricated spaces, eating in the identical themed restaurants that now sprout from the pavement in every corner of the continent to meet the spreading demand for anesthetic ease and familiarity. Each airport seems merely a new annex of the last airport, each new downtown an extension of the last: by lingering only in the hotels and conference centers where their meetings unfold, they've no need to subject themselves to the unruly otherness of the living locale.

Yet for those who have kept their animal senses awake—for those who venture beyond the made-to-order spaces, traveling more on foot or by bicycle than they do via the jet or the automobile—the journey from one ecosystem into another is precisely a journey from one state of mind into another, different, state. From one mode of awareness, flavored by salt and the glint of sunlight on waves, to an altered, inland awareness wherein the cries of those gulls become only a vague, half-remembered dream.

Transformations in the texture of the mind are not brought about only by journeying from one geography to another. Those who dwell steadily in a single terrain, whether by choice or necessity, also experience profound shifts in the collective awareness. The psychological qualities of any place steadily metamorphose as the powers that compose that place shift among themselves—as the dawning sun, for example, pours its first warmth across the rooftops—or as that place exchanges specific constituents with other places. The migratory route of certain birds may bring large flocks to settle for weeks or months in a particular bioregion; their arrival will alter the psyche of that realm even as those birds enter and partake of that very psyche. Such large-scale alterations in the collective sentience are especially obvious, while other, more continuous transformations go mostly unnoticed. Nonetheless, we can be sure that there *are* delicate changes unfolding in the local mindscape, changes that imperceptibly—but inevitably—affect our emotions, our thoughts, and our actions.

If the subtle intelligence of a place is inseparable from the medium of air that circulates invisibly within and between the inhabitants of that locale, it follows that some of the most dramatic modulations within the collective psyche are those that alter the sensuous quality of the atmosphere, changes that we commonly ascribe to "the weather." Transformations in the local weather often confound our conscious plans, sometimes curdling the unseen medium into a visible fog that slows our steps and clogs our thoughts, or congealing the invisible depth around us into a thicket of slanting raindrops. Changes in the weather transform the very feel of the world's presence, altering the medium of awareness in a manner that affects every breathing being in our vicinity. We sometimes refer to such weather phenomena, taken together, as "the elements," a phrase that suggests how basic, how primary, these powers are to the human organism. The ephemeral nature of such phenomena—the way such modulations in the atmosphere confuse the boundaries between the invisible and the visible, between inner and outer, between "subjective" and "objective"—ensures that the weather holds a curious position in the civilized world of modernity. We refer to it constantly; inquiring about or commenting upon the weather establishes the most basic ground upon which any social communication can proceed. Although it rarely occupies our full attention, the weather is always evident on the periphery of that attention, an ever-present reminder that the reality we inhabit is ultimately beyond our human control.

For the activity of the atmosphere remains the most ubiquitous, the most intractable, the most *enigmatic* of practical problems with which sedentary civilization has daily to grapple. Despite the best efforts of science and the most versatile technological advances, we seem unable to master this curious flux in which we're immersed. We rely on satellites to monitor its unruly behavior from outside, hoping to gain from such external data a rudimentary sense of its large-scale patterns, so we might better guess at its next moves.

But suppose we were to analyze this restless dimension *from within*—that is, from our own perspective as sentient animals thoroughly permeated by this flux? How then would we articulate its many modes of activity, its storms and its calms, its clarities and condensations as they resound in our organism and roll across the terrain? We would need a term that suggests the subjective quality of these elements, the way they alter the palpable mind of the place, transforming the awareness of all who live there.

For our own species, at least, it's clear that such changes in the ambient weather don't really *force* a change in our conscious thoughts, but rather alter the felt context—the somatic background within which thinking unfolds. From our own creaturely perspective, then, we might say that shifts in the weather are changes in the disposition of the land. Different atmospheric conditions—different kinds of weather—are, precisely, different *moods.*

Wind, rain, snow, fog, hail, open skies, heavy overcast—each element, or mood, articulates the invisible medium in a unique manner, sometimes rendering it (partly) visible to our eyes, or more insistently palpable upon our skin. Each affects the relation between our body and the living land in a specific way, altering the tenor of our reflections and the tonality of our dreams.

TORPOR

During the summer, near the coast, one sometimes wakes to a day that seems like any other, although as one goes through the motions of dressing and preparing breakfast one notices one's thoughts lagging behind, as though they have yet to fully separate themselves from the state of sleep. A slowness attends all one's cogitations—the newspaper today seems written with less zip, reporting the same old thing with the same stale phrases, and when you put it aside you wonder if a minute's rest on the couch would be in order before tackling the day's work. The immediate tasks to be accomplished are, after all, somewhat vague and unfocused; it's difficult to remember just what they are.

Only upon stepping outside, and surveying the world from your stoop, does the material cause of this mental lethargy become apparent. For the leafy trees, the electric wires, and the other houses are all bathed in a humid atmosphere that renders their outlines fuzzy and imprecise, while the mountains that usually rise from the far edge of town have dissolved, or are wholly shrouded, somehow, in the moisture-thick air. There's no cloud to be seen in the washed-out sky, only the too-big sun, hovering in the east, sweating like a spent tennis ball mouthed by too many dogs; its dull heat presses in from every direction.

LUCIDITY

Then there are those rare days, not entirely unknown in any region of the planet, that dawn with a clarity that muscles its way into every home and office, lending a crispness and cogency to almost every thought. One feels uncommonly good on such days, and others do, too; deliberations move forward with unaccustomed ease. Ambiguities resolve themselves—or render themselves more explicit, the choices more defined and clear-cut. There's a delicious radiance that seems to come from the things themselves, from even the tables and the plush rug, and when we step outside we can taste it in the air and the way a few fluffed clouds rest, almost motionless, in the crystal lens of the sky. How far our vision travels on such days! When we climb to the top of the street, we can see clear to the mountains that rise from the plain in the neighboring state. Long-term goals abruptly become evident; possibilities far in the future seem more accessible, lending perspective to the present. Hence planning goes more smoothly, with a marked absence of the usual friction—no sweat.

Although, to be sure, we're not always in sync with such felicitous weather—with the strangely clarified transparence that lifts the weight of the whole suburb on such unpredictable days, or that wraps the aspen branches outside our cabin with such a pellucid and form-fitting cloak of blue. Sometimes we're still carrying the strains and stresses of recent weeks, struggles that followed us into our dreams and now cling to our face and our feet, or perhaps we're still in the dank doldrums due to the wreck of a relationship we'd trusted our hearts to. These are the worst days for depression, when everyone else we meet moves so smoothly through the world. Even if we're off on our own, well away from the human hubbub, the despondence can be darker on such days, when it seems the stones and the singing sky and the blades of grass are all tuned to another frequency. There's an insistent and eager harmoniousness to things, an ease that we sense on the periphery—the hillside itself humming with pleasure for a whole afternoon—yet the mood cannot penetrate through the thick pellicle of our pain. The mismatch of the world with our own traumatized state feels distressing, even terrifying, shoving us deeper into the pit.

Of course, I am writing of these earthly elements, or moods, from an entirely human perspective. Indeed, I'm writing from the subjective perspective of a single human creature—myself. Nonetheless, I write with the knowledge that there cannot help but be some overlap between my direct, visceral experience and the felt experience of other persons—whose senses, after all, have much in common with my own. Moreover, I've confidence that my bodily experience is a variation, albeit in many cases a very *distant* variation, of what other, non-human, bodies may experience in the same locale in that season, at a similar moment of the day or night. For not only are our bodies kindred (all mammals, for instance, sharing a common ancestry), but also we are all of us, at the present moment, interdependent constituents of a common biosphere, each of us experiencing it from our own angle, and with our own specific capabilities, yet nonetheless all participant in the round life of the earth, and hence subject to the same large-scale flows, rhythms, and tensions that move across that wider life.

The world we inhabit is not, in this sense, a determinable set of objective processes. It is our larger flesh, a densely intertwined and improvisational tissue of experience. It is a sensitive sphere suspended in the solar wind, a round field of sentience sustained by the relationships between the myriad lives and sensibilities that compose it. We come to know more of this sphere not by detaching ourselves from our felt experience, but by inhabiting our bodily experience all the more richly and wakefully, feeling our way into deeper contact with other experiencing bodies, and hence with the wild, intercorporeal life of the earth itself.

STILLNESS

The computer keys fall and rise beneath my fingers, letters arranging themselves on the screen as I list the matters I must attend to in the next days. Drop off a clutch of letters at the post office. Compose a decent lecture for the conference on Thursday. Clean the stove and the oven. And since we're having guests, I'd better gather up this year's receipts from the various boxes and baskets and stand-alone piles where I've tossed them, and maybe organize them in a respectable fashion. The guests are my parents; they're arriving Wednesday—*tomorrow!*—to spend time with their two grandchildren. Tomorrow. How exactly do I expect to find time to craft a half-decent lecture for that conference? I'll have to come up with some sort of outline, and then just wing it. Meanwhile, my partner's been scowling at me all day because I forgot to fix the vacuum cleaner last week. (As if I've any idea how to even take that gizmo apart. When I fixed it last year it was sheer luck: the bulky part fell down the stairs while I was trying to pry it open, then miraculously starting working like new, so all of a sudden she decided I'm a mechanical wizard. Not that I don't enjoy the respect, but sooner or later she'll discover that I'm a quack, the last person one should trust with any tool more complicated than a shoehorn.) Anyway, I must remember—*remember!*—that the voice I reached at the power company said they'll indeed turn off the gas and electric if I don't get a check to them by Wednesday. Tomorrow. Too late to send it by mail. Maybe I can swing by there on the way to Hannah's dental appointment. After the dentist we'll have to head straight to the airport to pick up my folks; but why in Earth's name is she going to the dentist again—what kind of scam are these doctors running? If my daughter's teeth weren't perfectly fine, she couldn't bite her friend Finn so effectively; and if she comes home with one more coloring book from the waiting room, with all those grinning animal faces proclaiming how important it is to see the dentist every six months, I'm gonna sue the dentist for corrupting minors. Why does he need so many visits anyway—doesn't he have any friends?!

As the overwhelm hits, I feel a shot of acid in my stomach, and feel it register, too, as a fleeting wince on my face. Not good. This upcoming lecture has me on edge for some reason—maybe 'cause it's in my hometown, for once, and so various acquaintances will be there, expecting something dazzling. Christ.

Something catches my peripheral vision, and I turn toward the window. I feel my eyes widen in surprise: snowflakes! A great crowd of snowflakes floating down, a deep thicket of slowly tumbling white. How long has this been happening? I stand and stare for a few moments, then pull on a sweater and step out the door into a landscape transformed as if by a spell. My steps make no sound—the white blanket already plush upon the ground and layered in tufts upon the juniper and pine branches, as flakes drift down like loosened stars. A hundred of them swerve into my face, melting cold against my skin as I walk slowly through a world utterly transfigured by this silent grace cascading through every part of the space around me.

The surge and press of the week's worries have somehow vanished. When I try to call those concerns back to mind, I simply cannot find them behind the teeming multitude of slowly falling flakes. Past and future have dissolved, and I am held in the white eternity of a moment so beautiful it melts all my words. All weight has lifted; the innumerable downward trajectories have convinced my senses that I am floating, or rather rising slowly upward, and the ground itself rising beneath me—the earth and I ascending weightless through space.

A sound: the flutter of a bird's wings, and a small explosion of snow from a branch the bird launched from. Then, just silence. Not silence as an absence of sound, but as a fullness . . . as the very sound ten thousand snowflakes make as they meet the ground. A thick silence, muffling the whole valley, and for all I know the whole cosmos. I cannot imagine that any bird, any squirrel, any coyote or hare is not similarly held in the visible trance of this slowly cascading silence.

The clumps on the branches deepen . . .

The snow falls through the night, the porch light illuminating a charmed space through which powder floats steadily down. I turn it off before heading to bed, then step outside to inhale the darkness: now even the house, and the truck asleep in the driveway, have fallen under the spell.

By morning the snowfall has stopped. Yet the enchantment holds; when I step outside and snap my boots into my skis, there is a soft stillness everywhere. I glide between the trees and onto the dirt road, whose countless ruts have disappeared; unbroken goodness extends from the tips of my skis in every direction. There is a hushed purity to the world, and to awareness itself as I float across the snowy fields. The dentist will wait, and the power company will get its check when the roads are clear, whether today or tomorrow. Thursday's lecture is forming itself, easily, as I glide over the white expanse, my body writing its smooth script across the clean, unmarked pages.

Now and then a high limb releases its too-heavy mound of snow, and a spray of powder drifts down in sheets, glittering, scintillating, then vanishing into the clarified air.

WIND

Of all the elements, wind is the most versatile and protean, offering in each region a different range of personae, varying itself according to the season and often, too, according to the direction from whence it arrives. Even if we consider only a narrow range of its incarnations, we still find ourselves overwhelmed by their variety, and so must choose only a few examples from an outrageous array of styles:

Toward the tail end of a heavy winter, when a few days of unexpected warmth bring whiffs of spring, and a buried store of state-specific memories proper to that season send a few green shoots into one's consciousness, the winter will often reassert itself, swooping low at night to chill the walls of your house and repossess the snowy fields. When you step outside in the morning, the newly melted surface of the snow has now frozen solid and slick like a pane of glass. And gusting across that glass, a fine mist of crystals speeds past the trunks of trees, tinkling in eddies against the windows of the house as the main current of wind gallops through the valley. Boots slip along the frozen surface as you try to take a few steps, ears and face stung by the icy blast. Nothing in the landscape beckons or reaches out to you, for each bush or branch or telephone pole seems entirely focused on staying in place, every home holding fast with all its fingers to the ground beneath, each being doing its best to become an inconspicuous part of the ground, a mute lump or appendage of the earth, affording the wind nothing other than a smooth surface to glide past. Under the onslaught of the chill wind, each entity subsides into the anonymity of the earth, and even you, too, find your individuality subsumed into the rigor of standing solid against the frigid blasts, as your body makes itself into a smooth stone. Thought is stilled, all interior reflection dissolves, no memory apart from this ancient kinship and solidarity with the density of heartwood and metal and rock. The outward roar of wind forces one to find the blessed silence of stone at the heart of the mind. Anonymous, implacable, unperturbed—the biting cold of a winter wind returns one to one's unity with the bedrock.

Yet a wind of comparable velocity in the late spring can have a nearly opposite effect. As when after a long hike one ascends to a high pass from the eastern foothills and peers over into the valley beyond. A moist breeze is riding up the western slope, carrying fresh scents from the forests below, and clouds previously unseen are slowly massing on that side of the range. The wind becomes stronger, more insistent, and you realize that a storm is brewing; it's time to head down and find shelter. But something holds you on the pass. As the wind begins to rage, pouring over the crest and rushing down the boulder-strewn slope behind you, it tugs your hair back from your head and fills your cheeks when you open your mouth, whipping your unbuttoned shirt like a kite as an exuberance rises in your muscles. Laughing, crouching, and leaping in the wind, facing into it and feeling the first raindrops as you gulp from the charging gusts, imbibing the storm's energy, meeting its wildness with your own as you dance drenched like a grinning fool down the trail—a wild wind can return us to our own vitality more swiftly than any other element! And the needled trees swaying and tossing above us as we descend—are not they, too, caught up in something of the same mood? Not the giddiness, perhaps, but the exhilaration that lies beneath it, the way the wind challenges us in this season when the sap's already been rising in our veins, testing our flexibility, waking our limbs and our limberness, goading us each into our own animal abandon, our own muscular dance?

There are also the winds of autumn, those that whirl through the streets tearing the dry, ruddy-brown leaves from their moorings. Alive with the scents of fallen fruit and soil and smoke, the autumn wind teases our nostrils as it whooshes past, scattering the humped piles of carefully raked leaves, mingling their constituents with other leaves spiraling down from the branches. Soon the oaks, maples, and beeches stand denuded and exposed, their fractaled complexity silhouetted against the sky. Our bodies witness this gradual release of leaves, this stripping away of color from the gray, skeletal limbs, and cannot help but feel that the animating life of things is slipping off into the air—that the wind moaning in our ears is composed of innumerable spirits leaving their visible bodies behind. We feel enveloped by a rushing crowd of unseen essences—sighing, whooshing lives that reveal themselves to us only as fleeting smells, or by a momentary turbulence of dust and spinning leaves. The wind is haunted, alive. Only in this liminal season, before the onset of winter, does the wild psyche of the land assert itself so vividly that even the most rational persons find themselves lost, now and then, in the uncanny depths of the sensuous. Their animal senses awaken; the skin itself begins to breathe.

The disorientation is intensified by the autumn night—especially those bright, shadow-stalked nights when the howling wind hurls clouds across the face of the moon, loosening perception from its common constraints, since to our gazing eyes it seems not those glowing clouds that are racing south but rather the lustrous moon that is gliding smooth and swift toward the north, swimming upstream against the same current that buffets our legs and the gesturing branches.

For wind is moodiness personified, altering on a whim, recklessly transgressing the boundaries between places, between beings, between inner and outer worlds. The unruly poltergeist of our collective mental climate, wind, after all, is the ancient and ever-present source of the words "spirit" and "psyche." It is the sacred "ruach" of the ancient Hebrews, the invisible *rushing-spirit* that lends its life to the visible world; it is the Latin "anima," the soulful wind that *animates* all breathing beings (all *animals*); it is the Navajo "Nilch'i," the *Holy Wind* from whence all beings draw their awareness.

Indeed, whenever indigenous, tribal persons speak (often matter-of-factly) about "the spirits," we moderns mistakenly assume, in keeping with our own impoverished sense of matter, that they're alluding to a supernatural set of powers unrelated to the tangible earth. We come much closer to the shadowed savvy of our indigenous brothers and sisters, however, when we realize that the spirits they speak of have a great deal in common with the myriad gusts, breezes, and winds that influence life in any locale—like the wind that barrels along the river at dusk, chattering the willows, or the mist-laden breeze that flows down from the foothills on certain mornings, and those multiple whirlwinds that raise the dust on hot summer days, and the gentle zephyr that lingers above the night grasses, and the various messenger-winds that bring us knowledge of what the neighbors are cooking this evening. Or even the small but significant gusts that slip in and out of our nostrils as we lie sleeping. Modern folk pay little heed to these subtle invisibles, these elementals—indeed, we tend not to notice them at all, convinced that a breeze is nothing other than a mindless jostling of molecules. Our breathing bodies know otherwise. *But we will keep our bodies out of play; we will keep our thoughts aligned solely with what our complex instruments can measure. Until we have indisputable evidence to the contrary, we will assume that matter itself is utterly devoid of felt experience.*

In this manner we hoard and hold tight to our own awareness—like a frightened whirlwind spinning ever faster, trying to convince itself of its own autonomy, struggling to hold itself aloof from the ocean of air that surrounds it.

THUNDERSTORM

On a warm afternoon, new leaves creeping out of the just-opened buds, when the apricot trees shamelessly offer their blossoms to a thousand bees, one notices a faint rumble in the air. It dissolves back into the incessant whirring of bees, until there comes, sometime later, a similar trembling. The tremor is more felt than heard, a vibration noticed more by our bones and the trunks of the fruit trees than by our conscious reflections. Behind the branches, far off to the west, a darkness is gathering, a vague threat on the horizon. Yet now the irregular rumble once again, more audible, ominous. The rabbits are sniffing the air, hesitant. And how odd; what's become of all those bees? Now only a few stragglers are moving among the blossoms. Birdlife is more evident—several wingeds swooping between the trees, expending an unusual amount of energy. Everyone here is now feeling it: the background hush that's come over the land as the clouds thicken into a too-early dusk—the rabbits ducking into their digs—a deep quietude broken by the alarm call of a bird and a bit later by the thudding violence in the near distance. As though the sky is a skin that's stretched taut. Everyone is finding somewhere safe and hunkering down, tremulous, waiting.

Then, quietly, a soft breeze stirs the tips of the grasses, rippling the blades, spreading perhaps a kind of pleasure there among the green life, an eager anticipation very different from the threat vaguely sensed within one's muscles and the muscles of other animals.

And then without warning the air splits open: white fire tracing an impossibly erratic path between the sky and the hills opposite, a jagged gash burning itself into one's retinas and turning the entire landscape into a negative afterimage of itself, for an instant, before the shadowed darkness returns. Silence. And then the shattering sound of that splitting, the syncopated cracking, exploding in the skull and reverberating off the cliffs. The sound from which all other sounds must come. The Word at the origin of the world. And as the visible world settles back into itself, another bright flame rips haphazard through the gray, and soon the anticipated yet unprepared-for SHOUT!!! ruptures the air and shudders through the ground underfoot.

Nothing, no creature or stone or flake of paint on the wall, escapes the shattering imperative of the thunderbolt's shout—the way it undoes and re-creates us in a moment. No awake creature is distracted at that moment, no person remains lost in reverie or inward thought; all of us are gathered into the same electric present by the sudden violence of this exchange between the ground and the clouds, the passionate mad tension and static that reverberates through all of us in the valley this afternoon. This rage in the mind.

This passion now rising, it seems, in the branches of the ponderosas on the hillside opposite, and soon in the swaying limbs of the closer cottonwoods, and now in the roiling needles of junipers and piñons along the dirt road—some power is moving rapidly across the valley, a tumult of wind in the branches, and at last the rushing cool sound of

RAIN.

A few drops, at first, on my shoulder and nose, as I hear it begin to pelt the soil of the orchard. Then I am taken up within the cold thicket of drops, soaking first my clothes and then the smooth-skinned creature beneath those clothes, rolling off my nose and dripping off the apricot branches to pool among the grasses, spilling down my arms and gathering in the cuffs of my jeans.

The obvious effect triggered by the rain is release—a steady, dramatic release of tension, like held-back tears finally sliding across our cheeks.

Lightning still flashes through the downpour, and the stutter of thunder, but all this water drumming on my head eases the violence of that darker percussion, drawing attention back from the splintering tension in the sky to my own cool and shivering surfaces, and to the splashing patterns in the near puddles—returning awareness to the close-at-hand. Earlier, when the lightning first struck nearby, all attention was gripped by the present moment, yet that moment was a vast thing, opening onto the entirety of the clouded sky, including the whole lit-up span of the valley. A strong rain, however, rapidly shrinks the field of the present down to an intimate neighborhood extending only a few yards in any direction. The forest of droplets tumbling down all around me is not easily penetrated by my senses. Past and future seem utter abstractions, yesterday and tomorrow are far-off fictions; I am gripped in the slanting immediacy of water and mud and skin. I tilt my face upward, blinking, trying to follow individual drops as they fall toward me. Difficult. I give up and just open my mouth. The sensuous density of the present moment, and me inside it, drinking the rain.

I head into the house to peel off soaked clothes and towel myself dry. The rain beats an irregular staccato upon the roof. I stand at the window, staring out. Drops slam against different points on the pane, the impact bursting them into smaller droplets that slide waywardly down the glass, each droplet picking up others as it descends—every added straggler increasing the velocity of the drop, until they all pool along the bottom.

Even the interior of the house is transformed by the thrumming rain; objects here seem more awake to the things around them—the table, the lamp, couch, chairs, bookcases, and books all seem to have shed their distracting ties to the world outside and are now committed citizens of this small but commodious cosmos wholly isolated from the rest of the valley. The familiar bonds that these objects have with one another and with me are all heightened, somehow, by the pounding of water on the roof and the walls.

Later, after the rain has dissipated, I open the door onto a different world—a field of glistening, shiny surfaces, of beings quietly turning their inward focus back outward, as creatures poke noses out of burrows and a thrush swerves down to a puddle's edge and then hops in to splash its wings in the wet. Everything glints and gleams, everything radiates out of itself as a hundred scents rise from the soil and the fungus-ridden trunks, from insect egg cases and last year's leaves and the moist, matted fur of two squirrels chasing each other along the roof gutter. A tangle of essences drifts and mingles in the mind of this old orchard, each of us inhaling the flavor of everyone else, yielding a mood of openness and energetic ease as the clouds begin to part and the late afternoon sun calls wisps of steam from the grass.

But wait: Are we not simply projecting our own interior mood upon the outer landscape? And so making ourselves, once again, the source and center of the earthly world, the human hub around which nature revolves?

It is a key question, necessary as a check to our ingrained arrogance, and as a way of bending our attention, always, toward the odd *otherness* of things—holding our thoughts open to the unexpected and sometimes unnerving shock of the real. So are we merely projecting our emotional states onto the surroundings? Well, no—not if our manner of understanding and conceptualizing our various "interior" moods was originally borrowed from the moody, capricious earth itself. Not, that is, if our image of anger, and livid rage, has been borrowed, at least in part, from our ancestral, animal experience of thunderstorms, and the violence of sudden lightning. Not if our sense of emotional release has been fed not only by the flow of tears but also by our experience of rainfall, and if our concept of mental clarity is nourished by the visual transparence of the air and the open blue of the sky on those days of surpassingly low humidity. If our sense of inward confusion and muddledness is anciently and inextricably bound up with our outward experience of being wrapped in a fog—if our whole conceptualization of the emotional mood or "feel" of things is unavoidably entwined with metaphors of "atmospheres," "airs," and "climates"—then it is hardly projection to notice that it is not only human beings (and human-made spaces) that carry moods: that the living land in which we dwell, and in whose life we participate, has its own feeling-tone and style that vary throughout a day or a season.

In truth, it's likely that our solitary sense of inwardness (our experience of an interior mindscape to which we alone have access) is born of the forgetting, or sublimation, of a much more ancient interiority that was once our common birthright—the ancestral sense of the surrounding earthly cosmos as the voluminous *inside* of an immense Body, or Tent, or Temple. Before the invention of the telescope, the glimmering stars of the night sky appeared much closer than they do today. For the ancient Egyptians and Mesopotamians, the vault of the sky was considered the canopy of an enormous tent held up by the mountains that rise at the boundaries of the world. For the Haida people, a seafaring tribe of woodcarvers and wordsmiths who inhabit a rainswept archipelago off the west coast of what's now called Canada, to enter into one of their traditional cedar-plank houses was to situate oneself within a living analogue of the cosmos, for the universe itself was perceived as a huge house with holes in the planks where the stars gleamed through.

Similarly, a Navajo *hogan*—the traditional dwelling of the Dineh people of the American Southwest—is experienced as a microcosm of the enveloping house of the world. Indeed, in the absence of telescopes, some such conception of the cosmos as an immense interior or enclosure seems to have been common to a large majority of human cultures, as though a sense of the interiority of the surrounding world—this broad house of land and sky—is all but irresistible for the human creature.

The cosmology of Aristotle and Ptolemy, which held sway throughout Europe until the seventeenth century, was itself a refined instance of this same notion. It held that the universe was a nested series of concentric, crystalline spheres turning independent of one another around the central, solid sphere of the earth. Each of these transparent (and hence invisible) spheres carried one of the celestial bodies on its surface as it moved. Since to our unaided vision there were only seven independently moving bodies in the night sky, so in the pre-Copernican universe there were seven concentric, invisible spheres carrying the sun, the moon, and the visible planets, along with an eighth, outermost sphere carrying all the "fixed stars" as it turned. The sphere closest to us was that which carried the moon, followed by that bearing Mercury, then Venus, and then the sphere bearing the fiery Sun on its surface. Beyond that turned the sphere holding Mars, and then Jupiter, then Saturn (the outermost planet known to the naked eye), and finally the background sphere of fixed stars.

There was a great intimacy to this vision of the cosmos, with its invisible but ordered spheres enveloping the earth, cradling this world in their grand embrace. For all its complexity and observational refinement, it remained an extension of our ancestral, indigenous view of the universe as an immense enclosure. And so, when Copernicus and his followers wrecked this Aristotelian image of the cosmos, Western civilization suffered the dissolution of the last, long-standing version of that huge interior.

We can hardly imagine the visceral disorientation and sheer vertigo precipitated by that shift, as first the spheres holding the planets and then the outermost sphere of fixed stars abruptly dissolved into a boundless depth. Europeans soon found themselves adrift in a limitless space, a pure *outside*. Only in the wake of this dramatic disorientation, and the attendant loss of a collective interior, did there arise the modern conception of mind as a wholly *private* interior, and hence of each person as an autonomous, isolated individual.

Psychological qualities once felt to be proper to the surrounding terrain—feeling-tones, moods, the animating spirits-of-place known to reside in particular wetlands or forests—all lost their home with the dissolution of the enclosing, wombish character of the pre-Copernican cosmos. For unlike quantities, *qualities* are fluid properties arising from the internal, felt relations between beings. Such feeling-tones now had no place in the physical world—itself newly conceived as a set of objects that had no internal relation to one another. Nature was beginning to be experienced as a pure exterior—a world of external, mechanical relationships: a world of quantities.

It is only natural that psychological qualities fled from this open exteriority in the wake of the Copernican revolution, taking refuge in the private space now assumed to exist inside each individual. Feelings and moods are mercurial powers; they require at least a provisional sense of *enclosure* to hold them. Once they could no longer be contained by the sensuous cosmos, no longer held inside the curved embrace of the spheres, these fluid qualities quit the so-called outer world entirely, taking up residence within the new interiority of each person's "inner world." Henceforth they would be construed as *merely subjective* phenomena.

Yet there remains something exceedingly tenuous and unstable in this incarceration of felt qualities within the solitary precincts inside each person. For where, really, *is* this "inner" world? Does it exist somewhere inside our bodies? This seems unlikely, for whenever we open up a human body we find therein only a clutch of organs and tissues just as physically measurable as the things and objects outside the body. Perhaps it resides in the brain? If we cut into the brain, we discover a crowded mass of specialized cells, densely packed, with no breathing room between them. When we speak of our "inner world," however, we mean a spacious place where imagination dances and takes flight, an open field wherein thought has free rein, yet various ambiguous agencies (dimly sensed fears, tranquil yearnings, molten desires) vie for our attention, a realm with room for all kinds of adventurous forays. Clearly, then, we are referring to something other than the crammed thickness of our bodily interior. Where, really, is this expansive "inside" to which we allude, this spaciousness and depth wherein subjectively felt qualities have their home?

Despite a few thunderstorms last month, this land is dry—too dry. All around my home the piñon pines are desiccating and dying from the deepening drought. Most of the fish once endemic to the Rio Grande that waters this valley have now vanished; the river's flow usurped by too many industries, too many developments, too many golf courses. Sometimes clouds arrive, bringing hope of rain that hardly ever falls. Like today, a rare overcast afternoon in late spring—cloud cover stretching from the Sangre de Cristo peaks behind me to the Jemez Mountains on the western horizon, with nary a stitch of blue visible to the eye. The underside of the clouds is more defined than is common here in the high desert, the furrowed texture of that gray ceiling unusually clear. Here and there in the distance are small spots where a wispy, feathered blur slants down toward the ground: light rains falling on the far-off earth.

It is strange, this broad overcast extending its wings like a huge bird over the whole valley. The dense topography overhead seems as solid and palpable as the ground underfoot. And so the beings I meet as I walk between these two densities—rabbitbrush and sage, a red-shafted flicker testing with a few knocks the rotting shutter of an old window, a hare ducking under a sapless piñon, and the dry piñon snags as well—all seem curiously familiar today, as though we are co-conspirators in the same scheme, vital characters in the same broad story.

The dissolution of the concentric, crystalline spheres of the Aristotelian cosmos had opened the way for the very slow, gradual discovery of a new collective interior, less exalted than that earlier cosmos yet far more wondrous. After three and a half centuries spent charting and measuring material nature as though it were a pure exterior, we've at last begun to notice that the world we inhabit (from the ocean floor to the upper atmosphere) is alive. The feelings that move us—the frights and yearnings that color our days, the flights of fancy that sometimes seize us, the creativity that surges through us—all are born of the ongoing interchange between our life and the wider Life that surrounds us. They are no more ours than they are Eairth's. They blow through us, and often change us, but they are not our private possession, nor an exclusive property of our species. With the other animals, as with the crinkled lichens and the river-carved rocks, we're all implicated within this intimate and curiously infinite world, poised between the tactile landscape underfoot and the leaden sky overhead, between the floor and the ceiling, each of us crouching or tumbling or swooping within the same big interior. Inside the world.

THE SPEECH OF THINGS

(Language I)

The blades of my paddle slice the smooth skin of the water, first on one side and then on the other: *klishhh ... kloshhh ... klooshhh ... kloshhh ...* The rhythm matches the quiet pace of my breathing as I rock gently from side to side, gliding over the gleaming expanse of sky; the luminous vault overhead mirrored perfectly in the glassy surface. Tall, snowcapped mountains rise from the perimeter of this broad sea, and also seem to descend into it. In front of me, to the west, are the peaks of the Alexander Archipelago, the long cluster of islands off the southeast coast of Alaska; behind me are the glacier-hung peaks of the coast range. The liquid speech of the paddle sounds against the backdrop of a silence so vast it rings in my ears. The sky arcs over this world like the interior of a huge unstruck bell; the hanging sun is its tongue.

Between my kayak and those western slopes two smaller islands nestle close to one another. I am paddling toward them. I don't know the names of those islands, for I've not been in this region before. A breeze raises a pattern of ripples on the water's surface, then passes by: the mirror returns. *Klishhh ... kloshhh ... klishhh ...* Sometimes another, more rapid rhythm becomes audible as a pair of ducks materializes out of the near distance, flapping just above the water's surface. The thudding of their wings against the shallow layer of air swells in volume and then fades as their shapes dissolve back into the distance on the other side of my kayak.

The islands draw closer with each flex of my arms, widening their span and soon filling my gaze with green, gentling my ears with the liquid lapping of water against rocks. I sense vaguely that I am being watched. So I scan the rocky shore and the dense wall of forest above the high-tide line on each island, but can see no one. Only when a flash of white snags the corner of my eye do I notice the eagle perched high on a dead trunk jutting out from the coast of the more northerly island. Its lustrous head is cocked slightly—a single eye following the glints on my paddle blades. And perhaps the gleam off my glasses, as well, for when I turn my face toward it the bird launches with a few flaps of its huge wings, banks, and soars off through the passage between the two islands. I adjust my direction and follow it, gliding beneath the needled woods on either side. After a time I emerge from the channel; the echo of my paddle-strokes off the double wall of trees widens out and dissipates, giving way to a muffled sound drifting up from the south, a faint but dissonant clamor that rises and falls in intensity. Curious, I swerve the kayak to the left and begin paddling down the west coast of the southernmost island. When I round a spit of land, the noise gets louder, a low-pitched, polyphonic rumble that I cannot place at all. It fades to silence as I stroke across a broad bay, and then rises to my ears as I glide around another peninsula, although more intermittent now, and as I listen to this dark music I realize that it's an entirely organic cacophony, a crowd of rambunctious grunting tones vying with one another. As I cross the next bay it fades again. Only when the kayak slips around the next point and I see the long, rocky spit on the far side of the following cove—its jagged terraces and angled rocks bedecked with a jumble of sleek, brown humps—do I recognize that I'm entering the neighborhood of a large sea lion colony.

Oddly, the brown bodies opposite are mostly quiet as I come into view; a few grunts reach my ears as they negotiate places on the rocks. I can't make out any pups, and so this cannot be one of the rookeries where sea lions gather to breed and give birth, but must be one of their communal haul-out sites. A very *popular* haul-out site: I count over eighty adult sea lions as I paddle slowly across the cove, and know there must be many others hidden from view. But it's their immense bulk that startles me as I gaze through my binoculars. These are northern, or Steller, sea lions, far larger than their southern cousins; later I learn that the bulls can weigh up to 2,500 pounds, and can reach over eleven feet in length. I see some of them staring in my direction as I paddle. When I'm halfway across the cove, one such bull on a slab of a rock near the water raises himself up on his flippers, dips his head a couple of times, and begins roaring in a deep, guttural voice that resounds in the hollow of the kayak and reverberates in the cave of my skull. Soon two other large bulls lying on a ledge above the first raise their torsos and begin hollering as well, and within a few moments it seems every sea lion on that rocky outcrop is sounding its barbaric yawp over the waves. The raucous din is unnerving, and an upwelling of fear rises from the base of my spine. I lay down my paddle, and in an effort to quell the oncoming panic I do the only thing that I can think of, the single savvy act that might ease the tension in this encounter: I begin to sing.

This was a response to animal threat that I discovered some years earlier when, cross-country skiing along a snow-covered stream in the northern Rockies, I emerged from the woods into a small, frozen marshland—and abruptly found myself three ski-lengths away from a mother moose. She'd been feeding with her child among the low willows. The moose looked up as startled as I; she was facing me head-on, her nostrils flaring, her front legs taut, leaning forward. Her eyes were locked on my body, one ear listening toward me while the other was rotated backward, monitoring the movements of her calf. My senses were on high alert, yet somehow I wasn't frightened or even worried; I took a deep breath and then found myself offering a single, sustained mellifluous note, a musical call in the middle part of my range, holding its pitch and its volume for as long as I could muster. As my voice died away I already sensed the other's muscles relaxing. Drawing another breath, I sang out the same note again, relaxing my own body and pouring as much ease as I was able into the tone. Within a moment the moose leaned her head back down and casually began nibbling the willow tips. I sounded that liquid tone one last time, finally pushing off with my poles and slipping on past.

The simple appropriateness of what I'd done slowly made itself evident to my thinking mind as I glided through the woods. For the timbre of a human voice singing a single sustained note carries an abundance of information for those whose ears are tuned to such clues—information about the internal state of various organs in the singer's body, and the relative tension or ease in that person, the level of aggression or peaceful intent.

❦

And so, floating in my kayak, assailed by a chorus of bellowing grunts sounding from throats large enough, it seemed, to swallow me in a few gulps, I find myself singing back. Although not, this time, in a particularly mellifluous tone. If I had offered a gentle, calm note, the sea lions would never have heard me through the clamor of their own growling, and in any case I could never have generated such a soothing tone from within my already freaked-out organism. Instead, the musical tone that I utter forth is as loud and as guttural as I can manage, with my head thrown back in order to open my throat—a kind of low-pitched, gargling howl: "Aaarrrrrrggghhhh . . . , Aaaarrrggghhhh . . . , Aaaarrrrggghhhh . . ." I hold each guttural howl for as long as I can, finally pausing to draw a deep breath, at which point I notice, amazed, that the sea lions have stopped growling. I lower my head to look at them; they're now sniffing the air toward me, shoving one another to get a better glimpse of this large, brightly colored duck that can make such an ugly racket. My ears pick up the sound of fifty or sixty noses snorting and snorfling (and sometimes sneezing) as they sniff the breeze. My own nostrils can hardly sort the thickly mingled scents of salt spray and sea lion breath and the dense, floating beds of kelp as I take up the paddle and begin, like a fool, paddling closer. My own creaturely curiosity has gotten the better of my reason; I cannot help myself, enthralled by my proximity to these breathing bodies so weirdly akin to, and yet so different from, my own. The smell of them grows steadily stronger as I ease my kayak between the strands of kelp. When I get within about twenty-five feet of the rocks, that large male on the lower ledge—the same bull who initiated the alarm the first time—lifts his torso up on his flippers and starts bellowing. Straightaway a few others join in, and by the time I've laid the paddle across the kayak nearly all of the sea lions are hollering bloody murder. And so I am gulping air and mustering myself and about to launch into my own guttural harangue when, directly between me and the sea lions, the water's surface begins to bubble. Small bubbles at first, which soon give way to larger ones, and then a huge upwelling of water as, without any further warning, a gargantuan body blasts! through the surface into the sky—flying on outstretched wings that, as I stare wide-eyed, resolve themselves into the splayed pectoral fins of a humpback whale. The whale twists almost belly side up before its bulk crashes down, drenching me with spray and sending a huge wave rolling over the hull of the kayak, slamming the paddle against my life jacket and almost sweeping it away before I catch hold of its end and drag it back. In front of the kayak, the long, pleated folds of the humpback's underside are slipping slowly beneath the surface . . . and then the whale is gone.

I grab the paddle and desperately begin to back-paddle, thinking that the giant may try to capsize me, although after a few moments I realize that I've no idea what the whale is up to, or where in the depths it might now be. So I brace the paddle across the hull, gripping it tight with both hands, and simply wait. After a minute I hear the *pip, pip, pip* of tiny bubbles breaking, and by the time I locate them the water to my left begins boiling, and then upwelling, and before I can prepare myself that massive bulk explodes through the surface like a fever-mad hallucination—barely eight feet from the kayak—right side up this time and parallel to the boat although lunging in the opposite direction, immense pectoral fins dangling before it slams down. The swell catches my boat sideways and damn near flips me over, except that I counter-lean hard to the left, rocking back up in time to glimpse an incongruously small, almost human-like eye peering at me as it glides just above the waterline. The whale spouts, and a breeze blows its exhaled spray into my face, drenching my already sopped body, and then I'm overcome by the rousing stench of its breath. "Sewage-like," I think at first, but then it occurs to me, "What a blessing, to inhale the breath of a humpback whale!" The smell's intensity is jangling my neurons as the enormous apparition slides back down, leaving only a slim dorsal fin visible for a last moment before it vanishes beneath the surface.

I am left stunned, my entire body shaking in the kayak—the visual field trembling around me as I try to calm the tremor in my muscles. I feel as though the great god of the deep has just intervened between me and the sea lions, surfacing as a kind of warning, as if to say, *Not too close, mortal, to these kinfolk of mine!* Unable to quell the shaking, I lower my head to offer a mumbled prayer of thanks to these waters—but jerk my head back up as a loud SPLASH! sounds in my ears. My eyes widen in alarm. For the sea lions, apparently agitated by this visitation from the humpback god, are starting to dive off the rocks en masse. They're sliding down from the upper ledges and waddling over to the lowermost brink, where they're now plunging into the water in bunches, clusters of them tumbling into the brine and swiftly surfacing, and then surging—with their torsos half out of the water and with a holy clamor of guttural bellowing—straight toward *me*!

There is simply no way that I can escape their rapid advance: the fluid sea, after all, is *their* primary element, and not the customary milieu of this oafish stranger struggling to maneuver in his plastic, prosthetic body. I do not know by what wisdom, or folly, my animal organism chooses what to do next. Of course, there are not many options, and no time to think: my awareness can only look on in bewilderment as my arms fly up over my head and I begin, in the kayak, to dance. More precisely, my upraised, extended arms begin to sway conjointly from one side to the other, with my wrists and my splayed fingers arcing to the right, then to the left, then to the right, to the left, right, left, right . . .

As soon as I begin these contortions, the clamoring sea lions rear back in the water and fall silent, as their their heads begin swiveling from one side to the other, tracking my hands with their eyes. Astonishing! Seventy or eighty earnest mammalian faces twisting this way and then that way, this way and that, over and again. And all in perfect unison, like a half-submerged chorus line. After a couple minutes I drop my hands down to take up the paddle—but straightaway the sea lions start bellowing and surging forward. No! My hands fly back up and I resume the dance, my taut arms swaying left, then right, then left again as the whiskered crowd falls silent, their necks craning from side to side yet again, over and over.

My arms keep up their ritual, the kayak rocking this way and that. As I consider the situation, my happy relief at finding a way to save my skin gradually yields to a deepening dismay. For I can find no way out.

Whenever I even *start* to lower my hands the dark-eyed multitude lunges forward—so halting my dance is not an option. I examine my predicament from every possible angle, but cannot discern any exit strategy. And so I keep my arms high, inclining from one side to the other, smiling rather feebly at all these attentive, whiskered faces while the muscles in my upper arms grow more and more exhausted. After a long while the ache in my shoulders has become intolerable; I can no longer think. My right arm is giving out.

Slowly I bring that arm down while the left keeps up the rhythm. The sea lions, weaving from side to side, are now focused on the single, swaying metronome of my left arm. My right shoulder rests. An idea dawns. My gaze stayed fixed on the sea lions off in front of me as with my right fingers I begin groping around for the shaft of the paddle. On finding it I heft it slightly, balancing it as best I can in an underhand grip. Then, awkwardly, with my left arm rocking side to side above my head, I cross my right arm in front of my chest and begin rowing as best I can on the left. My right hand scrapes the unwieldy paddle against the left side of the kayak to get some traction. I do all this blindly, for my eyes are locked on the weaving faces of the sea lions, my left arm still swinging above my head. Slowly, arduously, my clumsy rowing manages to maneuver the kayak around the right flank of the floating mob. When most of the sea lions are off to the side, I bring down my left hand as well, clasping the shaft now with both sets of fingers, and begin paddling, *hard,* into the open water, without looking back. After seven or eight minutes I sneak a quick glance behind me: sure enough, a few sea lions are still trailing me, but at a respectful distance, and with little more than their noses above the surface . . .

Something in that charged encounter changed me. I notice it, sometimes, when I'm playing with my two children, or when the howling of coyotes wakes me in the middle of the night. My confrontation with the sea mammals brought home to me something crucial about language—something mightily different from what I'd learned at school and at college. I'd been taught that meaningful speech is that trait that most clearly distinguishes us humans from all the other animals. We have meaningful speech, while other creatures do not. But my unnerving meeting, in the wet, with the humpback and the mob of sea lions showed me otherwise. It made evident, in a way I could no longer ignore, that there exists a primary language that we two-leggeds share with other species.

When we speak of "language," we speak of an ability to communicate, a power to convey information across a thickness of space and time, a means whereby beings at some distance from one another nonetheless manage to apprise each other of their current feelings or thoughts. As humans, we rely upon a complex web of mostly discrete, spoken sounds to accomplish our communication, and so it's natural that we associate language with such verbal intercourse. Unfortunately, this association has led many to assume that language is an exclusive attribute of our species—we, after all, are the only creatures that use words—and to conclude that all other organisms are entirely bereft of meaningful speech. It is an exceedingly self-serving assumption.

Other animals, commonly possessed of senses far more acute than ours, may have much less need for a purely conventional set of signs to communicate with others of their species, or even to glean precise information from members of *other* species. My encounter with the sea creatures had initiated me into a layer of language much older, and deeper, than words. It was a dimension of expressive meanings that were directly felt by the body, a realm wherein the body *itself* speaks—by the tonality and rhythm of its sounds, by its gestures, even by the expressive potency of its poise. A near-catastrophic confrontation had plunged me into a space of earnest communication that unfolded entirely without words, a carnal zone of articulations broadly shared across species. It was a dimension wherein my verbal self was hardly present, but where an older, animal awareness came to the fore, responding spontaneously to the gestures of these other animals with hardly any interpolation by my "interior" thinking mind. It was rather as if my body itself was doing the thinking, trading vocal utterances and physical expressions back and forth with these other smooth-skinned and sentient creatures. Their flippers and fins were obviously shaped to a liquid medium very different from my own primary element, yet the most basic sensations of threat, or calm, or pleasure could still be swiftly exchanged—via the tautness or relaxation of various muscles, coupled with the tone of our uttered sounds—by virtue of our mutual existence as kinetic and sonorous beings inhabiting the same biosphere.

Sure, we were all mammals—the sea lions, the whale, and I—yet the sense I was left with was of a still more basic commonality or community of bodies, indeed of a communication shared as well with the waves shuddering under the kayak and splashing their speech upon the rocks. To the fully embodied animal *any* movement might be a gesture, and *any* sound may be a voice, a meaningful utterance of the world. And hence to my own creaturely flesh, as well, everything speaks!

Certain sounds that reach our ears convey the felt intent of other persons, while certain other, rumbling sounds bespeak a change in the weather. A rippling sequence of whistling tones expresses the exuberance felt by a thrush as the sun climbs above the horizon; other tones convey the dark magic of the night itself, speaking through the hissing tires on wet pavement.

Our human conversations are regularly influenced by this carnal layer of language, the apparent meaning of a friend's phrase altering with the pace of her speaking. The tenor of a spoken exchange may be transformed, without either of us noticing, when a break in the winter clouds allows the sun to spill its song over the muted hues of the city street where we stand, or by an abrupt and escalating argument of honking vehicles on the same avenue.

I began to notice this animal dimension in my own speaking—conscious now not only of the denotative meaning of my terms, but also of the gruff or giddy melody that steadily sounds through my phrases, and the dance enacted by my body as I speak—the open astonishment or the slumped surrender, the wary stealth or the lanky ease. Trying to articulate a fresh insight, I feel my way toward the precise phrase with the whole of my flesh, drawn toward certain terms by the way their texture beckons dimly to my senses, choosing my words by the way they fit the shape of that insight, or by the way they finally taste on my tongue as I intone them one after another. And the power of that spoken phrase to provoke insights in those around me will depend upon the timbre of my talking, the way it jives with the collective mood or merely jangles their ears.

Such was the linguistic dimension into which I was borne by that meeting with the lions of the sea—an initiation seared into my memory by the shock of being swamped by a humpback whale, and by the exchange of fetid breath with that wild intelligence. I now found myself more porous to other shapes, to smooth-surfaced desks and motley dogs, more aware of the conversation my animal body was carrying on with the other bodies around it, how it tensed in certain office buildings and loosened in dialogue with adobe walls. I noticed the skin on my skull tightening under the hum of fluorescent lights, and—once while cycling—felt my shoulder muscles open and expand as a red-tailed hawk took wing from a passing telephone pole. I heard more keenly how much my voice borrowed the rolling lilt of the person I was talking to, or took on the staccato stiffness of her syllables, and I noticed that she, too, was infected by the inflections of *my* voice, such that each conversation was also a kind of singing to one another, like two blackbirds trading riffs between the cattails—or like two humpbacks sending their eerie glissandos back and forth through the depths.

I've already spoken of my songful method for diminishing the threat posed by another, larger creature unexpectedly encountered in the backcountry—a simple way to convey that I intend no harm to the other. I should now report that my clumsy attempts at more nuanced communication with a wide range of animals, over many years of getting myself lost in the wild, have also made evident a uniquely efficacious way to bring one's specific intentions across to a member of another species. The technique—obvious, I know, yet only stumbled upon after much costly trial and error—consists in bringing your wandering attention entirely back to your own limber and sensitive body, becoming at ease with yourself and the slow rhythm of your breathing, and then just commencing to talk to the other animal in your mother tongue—in English or French or Inuktitut, or whatever language is really most comfortable. For if you speak honestly, then the audible modulations of your voice, along with the alterations in your visible musculature, and the olfactory emanations from your skin, will all be of a piece with the patterned meaning of your words, and so will readily convey something of your intent at a palpable, visceral level to the keen senses of the other animal.

Even when simply addressing a maple tree, or a boulder-strewn hillside, you can be sure—if you are honest, and so relaxed within your flesh—that there are sensate presences out and about that are affected by the sound and the scent and perhaps even the sight of your gestured intent, whether they be squirrels, or a swarm of termites chewing its way through the resonant hollow of a fallen trunk, whether a small, silent bat flapping erratically through the night air, or the airborne insects that the bat is hunting, or even the impressionable air itself, absorbing your chemical exhalations and registering in waves the sonorous timbre of your voice. And so your loquacious utterance is heard, or felt, or sensed—and it would be wrong to believe with certainty that you are not being understood. The material reverberation of your speaking spreads out from you and is taken up within the sensitive tissue of the place . . .

The activity that we commonly call "prayer" springs from just such a gesture, from the practice of directly addressing the animate surroundings. Prayer, in its most ancient and elemental sense, consists simply in speaking *to* things—to a maple grove, to a flock of crows, to the rising wind—rather than merely *about* things. As such, prayer is an everyday practice common to oral, indigenous peoples the world over. In the alphabetized West, however, we've shifted the *other* toward whom we direct such mindful speech away from the diverse beings that surround us to a single, all-powerful agency assumed to exist entirely *beyond* the evident world. Still, the quality of respectful attention that such address entails—the steady suspension of discursive thought and the imaginative participation with one's chosen interlocutor—is much the same. It is a practice that keeps one from straying too far from oneself in one's open honesty and integrity, a way of holding oneself in right relation to the other, whether that other is a God outside the world or the many-voiced world itself.

Nonetheless, the older, more primordial style of prayer sustains a very different stance toward the local terrain than that which resolutely directs itself toward a divinity beyond the world. While the latter feels the sensuous landscape as a finite and restricted realm relative to its transcendent source, the first experiences the sensible world as the *source of itself*—as a kind of ongoing transcendence wherein each sensible thing is steadily bodying forth its own active creativity and sentience.

To our indigenous ancestors, and to the many aboriginal peoples that still hold fast to their oral traditions, language is less a human possession than it is a property of the animate earth itself, an expressive, telluric power in which we, along with the coyotes and the crickets, all participate. Each creature enacts this expressive magic in its own manner, the honeybee with its waggle dance no less than a bellicose, harrumphing sea lion.

Nor is this power restricted solely to animals. The whispered hush of the uncut grasses at dawn, the plaintive moan of trunks rubbing against one another in the deep woods, or the laughter of birch leaves as the wind gusts through their branches all bear a thicket of many-layered meanings for those who carefully listen. In the Pacific Northwest I met a man who had schooled himself in the speech of needled evergreens; on a breezy day you could drive him, blindfolded, to any patch of coastal forest and place him, still blind, beneath a particular tree—after a few moments he would tell you, by listening, just what species of pine or spruce or fir stood above him (whether he stood beneath a Douglas fir or a grand fir, a Sitka spruce or a western red cedar). His ears were attuned, he said, to the different *dialects* of the trees.

When I tell others of this man's gift, overeducated folks often object to his turn of phrase, protesting the foolishness of alluding to the different "dialects" of the conifers as if there were actually a kind of spoken discourse in question. The rustling of needles, they point out, can hardly be considered the speech of a tree, since the sound is created not by the tree but only by the wind blowing *through* the tree. Curiously, these clever persons seem not to notice that it is demonstrably the same when *they* speak. We talk, after all, only by shaping the exhaled air that rushed into our lungs a moment earlier. Human speech, too, is really the wind moving through . . .

But meaningful speech cannot even be restricted to the audible dimension of sounds and sighs. The animate earth expresses itself in so many other ways. Last night while I lay sleeping the old apple tree in front of the house quietly broke into blossom, and so when, in the morning and still unaware, I stepped outside to stretch my limbs, I was stunned into silence by the sudden resplendence. The old tree was speaking to the space around it. Expressing itself, yes, and in the most persuasive of languages. The whole yard was listening, transformed by the satin eloquence of the petals. The spell quietly cast by the uttering forth of white blossoms was irrefutable and irresistible. (It has stayed with me all day, as a softness enfolding my thoughts, which is no doubt why I find myself writing of it now, late at night.)

❧

So language, from the perspective of the fully embodied human, seems as much an attribute of other animals and plants as of our own garrulous species. Yet, as we know from many of the traditional, indigenous peoples among us, this is still too restrictive: language accrues not only to those entities deemed "alive" by modern standards, but to *all* sensible phenomena. All things have the capacity for speech—all beings have the ability to communicate something of themselves to other beings. Indeed, what is *perception* if not the experience of this gregarious, communicative power of things, wherein even ostensibly "inert" objects radiate out of themselves, conveying their shapes, hues, and rhythms to other beings and to us, influencing and informing our breathing bodies though we stand far apart from those things? Not just animals and plants, then, but tumbling waterfalls and dry riverbeds, gusts of wind, compost piles and cumulus clouds, freshly painted houses (as well as houses abandoned and sometimes haunted), rusting automobiles, feathers, granitic cliffs and grains of sand, tax forms, dormant volcanoes, bays and bayous made wretched by pollutants, snowdrifts, shed antlers, diamonds, and daikon radishes, are all expressive, sometimes eloquent, and hence participant in the mystery of language. Our own chatter erupts in response to the abundant articulations of the world: human speech is simply our part of a much broader conversation.

It follows that the myriad things are also listening, or attending, to various signs and gestures around them. Indeed, when we are at ease in our animal flesh, we will sometimes feel that we are being listened to, or sensed, by the earthly surroundings. And so we take deeper care with our speaking, mindful that our sounds may carry more than a merely human meaning and resonance. This care—this full-bodied alertness—is the ancient, ancestral source of all word magic. It is the practice of attention to the uncanny power that lives in our spoken phrases to touch and sometimes transform the tenor of the world's unfolding.

❧

The sense of inhabiting an articulate landscape—of dwelling within a community of expressive presences that are also attentive, and listening, to the meanings that move between them—is common to indigenous, oral peoples on every continent. Like tribal people I've lived with elsewhere, most of my Pueblo friends here in the Southwest are curiously taciturn and reserved when it comes to verbal speech. (When I'm with them I become painfully aware of how prolix I can be, prattling on about this or that for minutes on end.) Their reticence is not due to any lack of facility with English, for when they do speak their phrases have an uncommon precision and potency. It is a consequence, rather, of their habitual expectation that spoken words are heard, or sensed, by the other presences that surround. They talk, then, only when they have good reason to, choosing their words with great care so as not to offend, or insult, the other beings that might be listening.

Here are some observations made by a member of the Mattole Indians (an Athabaskan tribe that traditionally hunted and fished along the Mattole and Bear rivers near the northern coast of California):

> The water watches you and has a definite attitude, favorable or otherwise, toward you. Do not speak just before a wave breaks. Do not speak to passing rough water in a stream. Do not look at water very long for any one time, unless you have been to this spot ten times or more. Then the water there is used to you and does not mind if you're looking at it. Older men can talk in the presence of the water because they have been around so long that the water knows them. Until

> the water at any spot does know you, however, it becomes very rough if you talk in its presence or look at it too long.[n]

These injunctions bespeak a remarkable etiquette, the careful deference and decorum to be observed when around water. While this decorum may at first seem ludicrous to modern sensibilities, notice: such an etiquette ensures that those who practice it will remain exquisitely attentive to the fluid ways of water—from the shifting eddies along the river to the tidal swells and rolling breaks along the coast. Such deportment, with its linguistic deference toward the fluid element, inculcates a steady respect for that element, ensuring that the community will not readily violate the health of the local waters, or the vitality of the watershed.

Few of us today feel any such restraints in our speaking. Human language, for us moderns, has swung in on itself, turning its back on the beings around us. Language is a human property, suitable only for communicating with other persons. We talk to people; we do not talk to the ground underfoot. We've largely forgotten the incantatory and invocational use of speech as a way of bringing ourselves into deeper rapport with the beings around us, or of calling the living land into resonance with us. It is a power we still brush up against whenever we use our words to bless and to curse, or to charm someone we're drawn to. But we wield such eloquence only to sway other people, and so we miss the greater magnetism, the gravitational power that lies within such speech. The beaver gliding across the pond, the fungus gripping a thick trunk, a boulder shattered by its tumble down a cliff or the rain splashing upon those granite fragments—we talk *about* such beings, about the weather and the weathered stones, but we do not talk *to* them. Entranced by the denotative power of words to define, to order, to *represent* the things around us, we've overlooked the songful dimension of

[n] G. W. Hewes, as quoted by Alfred Kroeber and Samuel Barrett in "Fishing Among the Indians of Northwest California," *University of California Anthropological Records* 21:1 (1960). I first found these lines in Freeman House's wonderful book, *Totem Salmon: Life Lessons from Another Species,* published by Beacon Press in 2000.

language so obvious to our oral ancestors. We've lost our ear for the music of language—for the rhythmic, melodic layer of speech by which earthly things overhear *us.*

How monotonous our speaking becomes when we speak only to ourselves! And how *insulting* to the other beings—to foraging black bears and twisted old cypresses—that no longer sense us talking to them, but only about them, as though they were not present in our world. As though the clear-cut mountainside and the flooding creek had no sensations of their own—as though they had no flesh by which to feel the vibration of our speaking. Small wonder that rivers and forests no longer compel our focus or our fierce devotion. For we talk about such entities only behind their backs, as though they were not participant in our lives.

Yet if we no longer call out to the moon slipping between the clouds, or whisper to the spider setting the silken struts of her web, well, then the numerous powers of this world will no longer address *us*—and if they still try, we will not likely hear them. They withdraw from our attentions, and soon refrain from encountering us when we're out wandering, or from visiting us in our dreams. We can no longer avail ourselves of their perspectives or their guidance, and our human affairs suffer as a result. We become ever more forgetful in our relations with the rest of the biosphere, an obliviousness that cuts us off from ourselves, and from our deepest sources of sustenance.

The propensity of our indigenous brothers and sisters to consult the animate earth around them, listening close to the land—carefully watching the patterned movements of other animals, attending to their diverse songs, signs, and gestures—all this is an obvious consequence of the expansive experience of language as a property that belongs to all things, and not solely to humankind. Given the near universality of this experience among native cultures, and given the fact that the abundant knowledge of indigenous peoples was traditionally transmitted *orally* rather than preserved in written form, we may suspect that literacy—reading and writing—

brings a dramatic transformation in the human experience of language.

As, indeed, it does. Our experience of linguistic meaning, however, is affected differently by different writing systems. Literate but non-alphabetic cultures, possessed of more imagistic writing systems—whether *hieroglyphic* (like that inscribed on walls and stelae in ancient Egypt, and the very different glyphic writing deployed by Mayan scribes prior to European conquest), or *ideographic* (like the written characters used throughout China)—all exhibit a wider reflective distance from nature than is evidenced by the more thoroughly oral cultures around them. Yet this distance is eased by the pictorial derivation of many of the glyphs or characters, which often borrow their shapes (or parts of their shapes) from elements in the surrounding landscape. In these scripts, highly stylized images of humans and human artifacts mingle with characters derived from the shapes of monkeys and serpents and trees, from the rising sun and falling rain. Those who read such scripts, then, are continually reminded of language's link to the more-than-human field of expressive nature. If the advent of writing grants a new visibility to human words—a somewhat static visibility that begins to enclose our gaze within the structured house of human language—nonetheless, many written characters in such scripts function as *windows* opening onto a living landscape that still speaks, onto a sensorial cosmos that still bears a kind of primary meaning.

Only when a more thoroughly *phonetic* system of writing spreads throughout a culture do its members come to doubt the expressive agency of other animals and of the animate earth. Only in the wake of the *alphabet* does language come to be experienced as an exclusively human power. The experiential shift can be attributed, in large part, to the way a phonetic script focuses our attention upon the specific sounds made by the human mouth. The written letters of an alphabet are no longer associated, by their stylized forms, with various entities and events in the surrounding earth. There is no indirect reference by the written characters to the sensuous world. Instead, the letters refer the reader solely back to his or her own mouth. Each letter, that is, is directly associated with a particular set of gestures, and sounds, to be made by the human tongue, lips, and palate.

Hence, instead of *windows* through which one might glimpse the wider landscape, the letters of an alphabet function more like *mirrors* reflecting the human back upon itself. Other animals—to say nothing of trees, mountains, and rivers—have no place in this new sign system, no expressive power in this new semiotic. Upon learning her ABCs, a human person can begin to dialogue entirely with her own signs, without any necessary mediation by the surrounding land.[Ω]

Of course, such a potent technology does not *force* its users to forget the animate, expressive power of earthly nature. But it does indeed make such forgetfulness possible, and for the first time.

Moreover, when reading an alphabetic text the reader finds himself in relation not only to a set of written injunctions, or to a clutch of compellingly written stories, but also to a voice, strangely like his own, that nonetheless seems to speak from an unchanging dimension apparently impervious to the growth and decay of bodily life. The alphabet, in other words, opens a new zone within human experience, a linguistic dimension that seems wholly unaffected by the flux of time.

It is only in relation to this powerful new magic (this technology that isolates and reifies human speech, disentangling the human tongue from the calls, cries, and whispers of the animate cosmos) that it becomes possible to intuit a humanlike God entirely *outside* the changing world, an omnipotent Voice to whom we, alone among all the animals, are beholden. Indeed, it is this very magic that catalyzed the emergence of monotheism, and the monotheistic faiths of the three "Peoples of the Book." That very phrase, "Peoples of the Book," should prompt us to recognize the

[Ω] This thesis is developed carefully and at length in my book *The Spell of the Sensuous: Perception and Language in a More-Than-Human World* (New York: Pantheon, 1996), chapters 4–7.

scribal nature of these three traditions, each of which has its codified origin in a different version of the alphabet (the Hebrew aleph-bet for the Torah, the Greek alphabet for the New Testament, the Arabic alphabet for the Koran).

Monotheism is a noble notion in many ways, and one that has brought abundant gifts to humankind. By loosening our enthrallment to the rest of animate nature, monotheism (always accompanied by its handmaiden, the phonetic alphabet) sparked a new and more detached curiosity in relation to our material surroundings, a new spirit of practical inquiry and experimentation, catalyzing an outpouring of creativity in the arts, in philosophy and the sciences, in the invention of new technologies. Yet the relative detachment from earthly reality inaugurated by monotheism seems to have hardened, today, into a cool and calloused imperviousness to the suffering of other creatures and the plight of the living land. Even as we discern the imminent danger to ourselves, we seem unable to locate any exit from the hall of mirrors, so thoroughly transfixed have we become by our own reflections.

Our intelligence struggles to *think* its way out of the mirrored labyrinth, but the actual exit is to be found only by turning aside, now and then, from the churning of thought, dropping beneath the spell of inner speech to listen into the wordless silence. Only by frequenting that depth, again and again, can our ears begin to remember the many voices that inhabit that silence, the swooping songs and purring rhythms and antler-smooth movements that articulate themselves in the eloquent realm beyond the words. Only thus do we remember ourselves to the deeper field of intelligence, to the windblown thinking that is not ours, upon which all our thought depends.[ה]

[ה] The steady subjection of experience to an unending, internal commentary is itself closely related to changes wrought by the written word—most specifically, the advent of silent reading. The ability to read silently is a fairly recent acquisition in the history of the alphabet. For many centuries, Greek and Latin texts were written with minimal or no punctuation, even without spaces between the words. As a result, readers had to sound out the text—reading aloud, or at the very least mumbling quietly to themselves—in order to distinguish particular terms and so to discover the precise meaning. It was

The ineffable and sacred One toward which all monotheisms direct themselves—the intuition of unity that long beckoned to our ancestors, the uncanny wholeness that seemed to whisper from behind the multiple local spirits and demons and gods—is still whispering even now, beckoning to us from beyond the monotonous hum and buzz of our worded thoughts, inviting us to free our senses from the verbal husk into which we've retreated. It calls to us from beyond the upholstered walls of these hollow

---

like encountering anendlessstringofletterswithoutknowingwhereanywordbeginsorends; reading aloud was necessary for most persons in order to disambiguate the visual text. (Ancient *Hebrew* texts, however, *did* employ spaces between words. Yet the reader of such texts *still* had to sound out the text orally, for the Semitic scripts had no dedicated letters for the vowel sounds! Since only the consonantal, skeletal structure of words was written on the parchment, the reader had to speak the text aloud in order to work out the precise worded meaning, enspiriting those bones on the page with his own breath in order to make them come alive and begin to speak.)

Beginning in the seventh century, various scribal innovations were gradually adopted by the monks copying texts in monasteries (the main method of producing books before the invention of the printing press), including the cutting-edge innovation of introducing spaces between the words. Aerating the text in this way made it possible for skilled readers to decipher the written text *without* sounding it out audibly, thereby inaugurating a practice of silent reading. As the practice of spacing and punctuating the written texts was taken up by more and more monasteries, the ability to read silently spread slowly throughout Europe. Yet it was not until the twelfth century that silent reading finally became commonplace among literate Europeans.

Learning to read silently is thus a fairly recent and hard-won accomplishment, one that forged a new association, in the human organism, between the visual focus and inner speech. Even as you read these printed lines, pay attention, if you can, to the phantom auditory sensation it entails—the silent "hearing" of a sequence of words within your head. The phantom "hearing" of those phrases is often lightning-fast, far faster than it would take to sound out the words audibly, and yet it is there. Now consider how similar that sensation is to the common experience of (verbal) thought. Notice how very alike is the felt sensation of thinking—the often involuntary awareness of worded thoughts rapidly unfurling within your skull—to the sensation of words arising inwardly as you move your gaze across these letters.

It is likely that the interior chatter of verbal thought, which for many people is as incessant as it is repetitive, was greatly amplified (if not inaugurated) by the popular advent of silent reading in the late Middle Ages. As our eyes, moving across the lines of text, learned to provoke an internal flow of words, a tight neurological coupling between the visual focus and inner speech arose in the brain. Given the growing emphasis upon the practice of reading for the cultivation of the self, it inevitably began to influence—and interfere with—other forms of seeing. Soon our visual focus, even as it roamed across the visible landscape, began to release a steady flood of verbal commentary that often had little, or nothing, to do with that terrain. Such is the unending interior monologue

caskets we've fashioned from the faiths of Abraham, Jesus, and Mohammed.

The mysterious voice toward which monotheism opens the heart is neither monotonous nor a monologue: it speaks through every utterance of breath and beak and branch, through the bugling of an elk and the liquid blink of a robin's eye. For many centuries it seemed to speak from a realm entirely beyond the tangible, since it promised a wholeness that far exceeded the scope of the terrain we perceived around us, a floating eternity into which our fleeting lives might return. Such a luminous unity, we thought, could only be of another, disincarnate dimension—for here in the physical, heaviness rules: all is leaden, a burden, and painfully bound to the ground. We could not imagine that solid flesh might float in space, nor that the rough landscape we saw around us might compose only the merest fraction, or glimpse, of an immense luminescent Body, a shape as perfect and boundless as a sphere, spinning.

We now know, however, that the tangible world is *itself* such an iridescent sphere turning silent among the stars, a round mystery whose life is utterly eternal relative to ours, from out of whose vastness our momentary lives are born, and into whose vastness our lives—like those of our ancestors, our enemies, and our children—all recede, like waves on the surface of the sea.

An eternity we thought was elsewhere now calls out to us from

---

that confounds so many contemporary persons—the "internal tape loop," or the incessant "roof-brain chatter," that Buddhist meditation seeks to dissolve back into the silence of present-moment awareness.

On the necessity of spaces between words for the cultural acquisition of silent reading, see *Space Between Words: The Origins of Silent Reading*, by Paul Henry Saenger (Stanford, Calif.: Stanford University Press, 1997). For the history of spacing and punctuation, see *Pause and Effect: An Introduction to the History of Punctuation in the West*, by M. B. Parkes (Surrey: Ashgate Publishing, 2008). Two other key works regarding the influence of such textual innovations upon the modern experience of the self are written by the necessary historian and cultural critic Ivan Illich. See his masterly book *In the Vineyard of the Text: A Commentary to Hugh's Didascalicon* (Chicago: University of Chicago Press, 1996), and his earlier book, co-written with Barry Sanders, *ABC: The Alphabetization of the Popular Mind* (New York: Vintage, 1989). For an abundance of evidence regarding the perceptual effects of the written word, see *The Spell of the Sensuous*.

every cleft in every stone, from every cloud and clump of dirt. To lend our ears to the dripping glaciers—to come awake to the voices of silence—is to be turned inside out, discovering to our astonishment that the wholeness and holiness we'd been dreaming our way toward has been holding us all along; that the secret and sacred One that moves behind all the many traditions is none other than this animate immensity that enfolds us, this spherical eternity, glimpsed at last in its unfathomable wholeness and complexity, in its sensitivity and its sentience.

THE DISCOURSE OF THE BIRDS

(Language II)

A sphere, suspended in the depths, its contour vague within the ink-black darkness, a black-on-black phantom so faint it may only exist in our imagination . . .

But watch: slowly, a thin margin of light rounds one edge of the sphere, outlining the orb's curve against the dark like a golden bow stretched taut. The light's edge is now gliding almost imperceptibly across the disk, leaving a radiance in its wake: the bright bow slowly thickening into a crescent of luminous blue. Soon other colors (white, brown, green) shape themselves within the blue as the shining frontier advances, until fully half of the orb's visage is illuminated, then three-fourths, and then the entire face gleams perfectly round within the fathomless deep.

Although already that face feels uncentered, unbalanced. Why? Look carefully: a new margin, it seems, is now following the first, an edge of obsidian darkness sliding ever so slowly across the face, erasing its features as it spreads. And listen: a faint sound like a whisper seems to accompany the frontier of shadow as it proceeds. Can you hear it? Move in closer, listening . . . Yes, the leading edge of the dark is indeed an audible as well as a visible line, a whisper rolling slowly across the sphere like a wave, drowning its colors in the anonymity of night.

When the darkness reaches the farthest bound of the sphere, another audible wave slips in at the first side and begins its own journey, illuminating the orb as it advances, releasing the blue and the swirling white. The whispering, whistling sound that it makes as it moves seems identical to the hush of the previous, darker wave—a hypnotic, breathy music. Let's swoop in closer still, to hear . . .

Ahh, now that moving frontier sounds like a many-layered rustling, an audible curtain woven from intertwined strands of rushing wind and watery tones, some rising and then falling, others dipping and lifting, ethereal. As we drop in to the sphere—its half-shadowed face expanding to fill our gaze—the rushing exuberance of this strange music pours over us, its tangled strands now distinguishing themselves as whirring, fluted phrases and whistling rhythms, as the buzzy trills of towhees and the syncopated quaver of song sparrows, as winter wren's fevered twitterings and meadowlark's fluid gurglings, the loopy gossip of solitaires against a woodpecker's hammerings, loud nagging of jays and nasal warbling of house finches: all composing a dense membrane of speech sweeping slowly over the land, a sonic edge carrying warmth in its wake, an audible line with sections that fall almost silent as they pass over large bodies of water, over bays and seas, then grow louder as they approach dry terrain. The advancing line swells with plaintive cries of gulls and nasal whistles of sandpipers, avocets, and the guttural grunts of cormorants; with the singsong of curlews, goose honks, and the thrum of hummingbirds swerving between blossoms, now the seesawing lilt of robins, hoarse phrases of tanagers, a bunting's mad rhythms, here the eerily ascending arpeggios of a Swainson's thrush, there the cascasding arpeggios of a canyon wren, raucous banter of ravens (punctuated with clicks and fluid boinks), and then a loon's mournful yodeling—each utterance with its own place, its own tonal niche in the rolling wave of discourse, within the swelling speech of the dawn as it breaks and spills the morning's light over the land. And as the wave passes and moves on, the cacophonous chorus of each place quiets down, the individual singers giving themselves over to the day and its necessary doings.

To be sure, talk is still exchanged throughout the fields and the fragmented forests; across even the treeless shantytowns and the suburban backyards. There are the necessary check-ins, gossip, even an occasional bursting into song, yet it's far more subdued and desultory than the dawn's exaltation—no compulsion to expend such energy now when it's needed elsewhere (finding food, raising the kids, avoiding predators). Only much later, as shadows lengthen and the light thickens to yellow and honey gold on the sides of houses, does the impulse begin to rise once again in the throat, drawn forth by the slowly sinking sun. And so from the fences, and telephone wires, and the upper branches of cottonwoods, a soaring crowd of voices lights the late-afternoon air, an interlaced paean of ebullience and earnest pleasure swelling with the precise approach of evening's edge, as though it were the very voice of that edge. Which, of course, it is.

Until, as dusk dims into night, the choral abundance fades into a quietude much deeper than the muted talk of the day, a cool hammock of silence slung between the soft scratching of crickets and the far-off hoot of a great horned owl.

Such is the strange world we inhabit: an immense sphere around whose surface two long lines of birdsong are steadily sweeping—always opposite one another, two breaking waves of vocal exuberance rolling ceaselessly around the planet.

❧

The wind is rising. I'm walking in a coastal estuary, the outflow of a small tidal river on the north coast of Long Island, near the city of New York. Over the last hour the ebbing tide has revealed more and more curving sandbars as the river's waters slither among them, its central channel now accompanied on either side by smaller streams, rills and runnels carving delicate, rippling patterns in the sand. I wade through these various currents, heading toward the bay into whose waters this now-fragmented, many-fingered river is releasing itself. Above me huge, storm-laden clouds are steadily massing, blown landward by the wind out of the north, their leaden grays and maroons thickening, eclipsing the last patch of blue as they merge into a single, roiling topography overhead. The winds are now hissing the clumped grasses and wailing through a stand of limber trees rooted in a spit of land off to my right, the trees bending so low it seems they're about to break. But they don't yet, although the squall is gusting hard against my chest, shoving me backward again and again before I stumble forward. Now my face is being pelted with drops, whether from surface spray kicked up by the wind or rain finally falling I can't tell, as the moan of the air rushing past my ears rises and falls. Soon a mist blurring the sea in front of me makes evident that it's really raining. Already drenched, I keep leaning into the blustering surge, making my way through the tumult of unseen powers howling and whooshing and tearing at my coat.

And then, in the thick of this tempest, I hear several people serenely conversing. The measured calm of their voices somehow calms me down as well, though I cannot make out what they are saying. But where are they? Turning this way and that, peering through the gusting mists, I can see no one. Yet they sound so close at hand! Somehow these voices are being carried a long distance by the wind. Gazing around again, I assure myself that there's nobody off in front of me, or behind me, or visible on either side. And then I look up . . . and abruptly recognize them, there, about seven feet above and a bit in front of me: five ducks flying into the wind, or at least trying to. They're making no headway whatsoever, the gusts blasting so hard against them that with all their strenuous flapping they are unable to advance, but are basically held motionless in the surge. The five of them are arrayed in a haphazard line about twelve feet across. Every now and then one falls back and plunges toward another point in the wall of wind, hoping to find a soft spot. But to no avail. It seems obvious that the group of them had flown up behind me, and were heading out toward the open water when they ran into this bank of wind, an invisible edge held in place by the headlands around us. Still, they persist, and as they strive against the surge they call to one another in brief, oddly calm-sounding quacks; it was these phrases, reaching my ears through the maelstrom of wind, that at first sounded so like snatches of human conversation.

I listen to them now, as I watch their relentless attempts to break through the impasse, with sometimes one and then another making a brief foray or trying a new maneuver, and after a while their quacking speech begins to make new sense to my listening self. As if my watching eyes had clued in my listening ears, and so ears and eyes had abruptly merged into a single keen sense—for now I can *feel* those five birds much more clearly, can sense their strangely different intelligence just above me, there, as they wrestle the wind. I can now sense the clear intent in their staccato voices as they speak, as what had sounded like a single repeated utterance now varies subtly in rhythm and volume according to whether the duck speaking is testing the wind with its muscles or simply holding the status quo. Each voice alters its feel when the speaker is blown off course by the gusts, each duck using its quacks to inform the others about the state of the blast just in front of it while also apprising them of its precise location at that moment (since they're unable to glance around without ceding ground to the wind), each also replying and reassuring the others, so that a whole array of nuanced meanings is passing back and forth above me.

The convergence between my listening ears and my gazing eyes has brought me much deeper into my animal body and my body's world. It is now indelibly evident to me that these birds are not just beings of instinct—at least no more than I am—but are fully awake entities earnestly engaged in the thick of the present moment.

The extra strength of the wind at that spot, blocking the ducks from flying farther, is an interference pattern, I suppose, between two moving masses of air, although not being a bird, I've not had to pay attention to such flows in the unseen atmosphere, and so have scant savvy regarding the wind's vicissitudes. It now occurs to me that the wingeds must know from daily experience a very great deal about the fluid structures and shifting topologies in the unseen medium of air. Probably many mammals—at least those who walk on all four of their legs, and so have not curtailed their use of smell—know all sorts of things about the shapes that smells take as they propagate on the wind: a potential mate's distinctive trace spreading as an airborne gradient that alters with the breeze, or the phantom presence of one's prey wafting as a ribbon of scent just above the soil. And winged insects, of course, chart their course by such gradients. But it is the birds—in their outrageous, many-feathered diversity—who have the greatest and furthest-ranging knowledge of the large-scale voluminous shapes that form and metamorphose within the invisible medium. It's they, after all, who ride those shifting flows, they whose muscles daily navigate the unseen hints and whirling notions that ceaselessly arise, interact, and dissipate in the subtle awareness of the planet.

How easy it is for inherited concepts to stifle our senses! So often we assume that other animals are not conscious—that birds, for example, lack real intelligence, since their brains (or their "brain-body ratios") are so much smaller than ours. "Birdbrains," we say—a facile insult. Intelligence, we assume, is a strictly centralized phenomenon, mistaking our distinctively human form of intelligence for intelligence itself. As a mammal who long ago learned to balance on its hind legs, freeing its forepaws to manipulate objects, we specialize in a kind of curiosity that looks at things from different directions, turning them this way and that—a highly visual pondering that seems to unfold somewhere behind our eyes. It is a propensity for detachment that, as this book has already suggested, has been greatly intensified in the current epoch. An expanding complex of technologies now mediate between our body and the earthly elements; instead of navigating the elements directly, we're accustomed to adjusting the switches on those technologies, using our intelligence to maneuver among a determinate set of abstract parameters rather than to improvise our way through an indeterminate and ever-shifting material field. When driving our cars, or while sitting in airplanes, we're accustomed to fairly mechanical operations that require little strenuous effort. Whether sequestered in our offices tickling the keys on a keyboard, or simply turning the pages of a good book in bed, we spend much of our time deploying a very rarefied form of intelligence, manipulating abstract symbols while our muscled body is mostly inert. Hence thinking, for us, seems to have little bearing on our carnal life; it often seems entirely independent of our body and our bodily relation to the biosphere. If our reflections *do* result in a change of stance or an alteration in our actions, they still seem to issue as directives from a centralized thinker—or self—oddly independent of our materiality, a floating locus of awareness situated somewhere within our heads.

Other animals, in a constant and mostly unmediated relation with their sensory surroundings, *think with the whole of their bodies.* A nuanced creativity is necessary to orient and forage in a world of ever-changing forces. Equipped with proclivities and patterned behaviors genetically inherited from its ancestors, each wild creature must nonetheless adapt such propensities to the elemental particulars of the place and moment where it finds itself—from an unexpected absence of water in the usual watering hole, to a sudden abundance of its favorite food. No matter how precise are the instructions tucked into its chromosomes, they can hardly have encoded, in advance, the exact topology of the present moment. And hence a modicum of creative engagement in its immediate circumstance is simply unavoidable for any organism that moves (whether an elephant or an amoeba). It is unavoidable for any organism that must orient itself, now and then, to the rest of a steadily transforming reality. A spider sets about mending her web after it's torn by a pounding rain, adding only those strands necessary to patch the precise dimensions of the hole. A monkey skillfully adjusts the strength of its arboreal leaps to match the distances of the various branches that loom up one after another. A foraging bull elk, stymied in his attempt to rejoin the herd by a newly flooding river, assesses his chances; will he plunge in to ford the swollen waters, or wait out the tumult?

Some may say that such decisions are not thoughts, that the animal is not aware of these choices. Yet it is clear that SOMETHING is aware in the present moment, monitoring the terrain and responding accordingly. It is true that we need not attribute such choices to a *separable* self within the animal, or to a thinking that unfolds entirely in some autonomous part of the creature's brain before issuing its clear commands to the limbs. Never having separated their sentience from their sensate bodies—having little reason to sequester their intelligence in a separate region of their skull where it might dialogue steadily with itself—many undomesticated animals, when awake, move in a fairly constant dialogue not with themselves but with their surroundings. Here it is not an isolated mind but rather *the sensate, muscled body itself that is doing the thinking,* its diverse senses and its flexing limbs playing off one another as it feels out fresh solutions to problems posed, adjusting old habits (and ancestral patterns) to present circumstances.

This kind of *distributed sentience,* this intelligence in the limbs, is especially keen in birds of flight. Unlike most creatures of the ground, who must traverse an opaque surface of only two-plus dimensions as we make our way through the world, a soaring bird continually adjusts minute muscles in its wings to navigate an omnidimensional plenum of currents and interference patterns that alter from moment to moment—an unseeable flux compounded of gusting winds and whirling eddies, of blasts and updrafts and sudden calms, of storm fronts, temperature gradients, and countless other temperamental vectors and flows that may invisibly and at any moment impinge upon your feathered trajectory—whether from in front or above or below, shoving you from one side or the other or from several directions all at once. Flying is an uninterrupted improvisation with an unseen and wildly metamorphic partner. In order to stay aloft and on course (whether toward a certain stand of trees or an unsuspecting rodent moving among the grasses), tactile and kinesthetic sensations registering changes in the enfolding currents must translate instantly into muscular adjustments, diverse sinews flexing and extending to counter such perturbations even as those currents keep shifting, in a dynamic, rolling exchange between your wing muscles and the rushing muscle of the wind itself. It's a kinetic conversation in the uttermost thick of the present moment. A barn swallow swooping, dipping, and banking for insects; a kestrel hovering directly above a field mouse; a crowd of crows mobbing a hawk; the bald eagle I once watched diving, talons first, toward a young loon; a gull kiting in the wind just above me, keeping perfect pace with the ferry I'm riding—whatever other styles of savvy these creatures may or may not display, all of them, while aloft, are thinking exquisitely along the whole length of their extended limbs. It is a brilliance we're ill equipped to notice if we associate smartness only with our own very centralized style of cogitation. When we disparage the intelligence of birds, or the size of their brains, we miss that flight itself is a kind of thinking, a gliding within the mind, a grace we humans rarely attain in our contemplations (although if we're following a falcon with our focus, we sometimes find our thoughts soaring as well).

Of course, we humans also think with our muscled limbs—less elaborately than the wingeds, but still: our legs steadily adjust their stride to match the steepness of the slope, our ankles flexing to meet the exposed roots and jumbled rocks as we walk. It's an ongoing and attentive response to the unpredictable nuance of the present moment, a corporeal decision-making that underlies all our abstract reflections. Nonetheless, we've learned to associate thought only with the latter—with the very verbal (and sometimes numerical) cogitations coursing within our heads—and so have taken ourselves out of rapport with the other animals.

Walking through the forest, we often fail to register the vocal sounds of other animals, the whistles of squirrels and the intermittent calls of various birds, because although our bodies are in the forest, our verbal thoughts are commonly elsewhere. As Thoreau chides himself, "What business have I in the woods if I am thinking of something out of the woods?"[Ω] What business indeed! As soon as I call my errant spirit back home to its senses, my animal organism awakens from its slumber. Now the snap of a far-off twig brings a new alertness to my listening, as the hypnotic humming of insects and the dark squeak of two trunks rubbing against one another yields a keen awareness of my proximity to lives being lived at different scales from my own.

Here in the forest, all is body language. Tall spruces, orb-weaving spiders, a chipmunk poised on a fallen trunk rapidly gnawing something held in its forepaws, even the Jackson Pollock outbreak of bright lichens on a rock outcropping—all of these breathing beings are *bodies,* distant variants of my own flesh, as indeed my body is a distant echo of theirs. If each also has its sensations, its own experience of the world around it (which appears likely, since each responds appropriately to its context), it seems obvious that their experience is as weirdly different from my experience as their bodies differ from mine. We are almost wholly alien to one another.

Yet each organism in these woods seems to express itself directly, without the mediation of symbols or sentences. Hence the tension expressed by the sounds or movements of another creature will sometimes trigger a resonance in my own flesh. I've no doubt that my empathic sensations are dramatically different from those actually felt by the skittish deer or the squirrel, yet with regard to such basic experiences as fear, pain, and pleasure, it seems silly to assume that our feelings are entirely incommensurable. There is a subtle entanglement and confusion between all beings of the earth, a consequence not only of our common ancestry, and the cellular similarities of our makeup, but also of our subjection to variant aspects of the same whirling world.

[Ω] From the essay "Walking," in Henry David Thoreau, *The Natural History Essays* (Salt Lake City: Peregrine Smith Books, 1980), p. 99.

Taking a cue from my friend Jon Young, a remarkably gifted tracker trained in several indigenous traditions, I've begun to tune my ears to the discourse of the local birds. Jon pointed out to me that there are five basic phrases in the vocabulary of most perching birds—a simplification, perhaps, but one that has enabled my listening to gain a first access to the language of the winged folk. The five elemental phrases that Jon identified among the perching birds (or passerines) are these: the song itself, the companion call, begging cries, male-to-male aggression calls, and the alarm call.

The *song,* a melodic string of tones and trills heard especially in spring and summer, is often particular to the males of a species. The song is a prime way of attracting females, and also seems to function as a territorial display, proclaiming one's space and saying to others: keep out. Yet the songs of many species have also—to my ears at least—an exuberant and often celebratory quality unacknowledged by those who insist upon a strictly functional account of their intent. There's often a palpable feel of contentment in the song (even, at times, a sense of real pleasure in the song's production). It is this quality—the contented feeling-tone of the song—that instantly indicates, to a savvy listener, that there's no overt source of distress in the vicinity of the singer. If a bird is sounding its song, we can be reasonably sure that there's no evident danger lurking about.

The *companion call* is rarely indicated in my birding field guides. The call is commonly uttered by both the female and the male of a mated pair, usually in an alternating pattern. It seems a way of staying in close, auditory contact—a kind of *checking in* with one another—a brief chirping back and forth that lets each bird know the other's whereabouts while both are foraging. Slight variations in the call may serve to indicate that the caller has found a good food source—"come on over"—or any number of other nuances. Whenever we hear the companion calls of one species flowing back and forth in a regular rhythm, it's a clear indication that the birds are in a relaxed or "baseline" state, a condition of ease that expends little unnecessary energy. If the rhythm is interrupted—if one mate stops replying—then the other will call again in an irregular pattern, sometimes raising the volume. This interruption in the normal pattern is instantly noticed by other wildlife in the area (not only by birds but by squirrels, deer, fox, and others), alerting them to the possible presence of a predator or intruder. The alertness rapidly deepens if there's no response for a long stretch. But if the first bird finally responds, then the rhythmic pattern of chirps will reassert itself, and the neighboring animals all relax.

The *begging calls,* usually heard in late spring or early summer, are made by fledglings calling for food, and cue us to look for the diligent behavior of the parents (darting hither and thither, hunting for the desired nourishment, and winging it swiftly to their young). Unlike the song or the companion call, begging cries don't necessarily indicate a state of safety, since fledglings often aren't sharp-witted enough to shut up when there's a raptor or other predator nearby.

Aggression calls generally arise when a male songbird flies into the territory of another male from the same species. This call is often sounded on the wing, as one male tries to chase the other out of his space. The aggressive intent can usually be heard in the sound itself, and is especially obvious when one glimpses two birds with the same intense cry chasing one another. A good deal of energy is expended, but if we look around, we'll see that other birds pay scant attention to the goings-on.

The *alarm call* is uttered whenever danger is sensed—when, for instance, there's a potential predator in the vicinity. It's both an expression of dismay and a stark warning to others, and it usually varies in volume with the degree of sensed danger. This call is the clearest sign of felt peril, and a clear departure from the baseline state. Upon hearing an alarm call, even birds from other, neighboring species halt whatever they're up to, opening their attention toward the possible threat. Other animals, too, commonly stop and take notice. Wild mammals are generally familiar with the alarms of the local songbirds; hence, whenever such a call is uttered, the entire neighborhood is drawn to attention.

Avian alarms are also often triggered by us humans as we approach—although, caught up in our own musings, we're usually oblivious to the wary commotion we set off as we wander the woods. If we're alert to such cries, then the trajectory of another person walking among the trees can readily be tracked by our listening ears as a spreading ring of alarm accompanying that person as she moves, radiating outward in concentric circles from her presence.

Since many songbirds are ground feeders, they fly upward into the trees when they first sound the alarm, usually alighting (and this is crucial) on a branch *just high enough* to be out of harm's way. So if we hear a series of alarm calls moving slowly through the forest, and the calls seem to issue from a height around seven or eight feet above the ground, it's very likely that a human is setting off the disturbance. But if those alarms are sounding from a height of only three or four feet, it's probable that a smaller, ground-dwelling predator is moving through the woods—a fox, perhaps, or a weasel.

If, however, the alarms all seem to sound from the upper canopy (from those birds, like chickadees and warblers, that hang out in the high branches), then an aerial predator is likely soaring or swooping nearby, and if we cast our eyes upward we've a chance of glimpsing the raptor that is causing all the hubbub.

Because the calls of birds often carry quite far through the tangle of branches and trunks that so easily confounds our gaze, by learning to distinguish the various vocalizations of the local songbirds—and by noticing whether a call gives evidence of a relaxed mood or expresses a state of alarm—we can suss out much of what's happening at a fair distance from us in the woods. If, by consulting the neighbors and a few pocket field guides, we familiarize ourselves with the most common predators in our area, then by listening to hear the general elevation from which alarm calls are sounding—and by following, with our own animal ears, the passage of such calls as they propagate—we can glean the possible identity of the intruder, and even trace the trajectory of that creature as it moves within the forest. Such simple skills, honed by continued practice, clue us in to the activity of other animals, enabling us to move toward them and sometimes glimpse them without triggering such alarms ourselves.

As we begin to tune in, there's an eerieness that dawns when we realize that *most other animals are also listening closely to the discourse of the birds, and have been doing so all their lives.* Since birds are by far the most agile creatures within the forest, able to swoop from place to place and to gaze around as they fly, they naturally know a great deal more about who's presently afoot in the woods, and where, than the other animals. Their spoken utterances, each species with its distinct phrases (varying in tone in different circumstances), serve to keep the entire forest well apprised of what's occurring within its depths. By familiarizing ourselves with the most basic terms of their language, we avail ourselves of a detailed outlet for "the news" already relied upon by numerous creatures.

Perhaps for this reason, the sacred language regularly attributed by tribal peoples to their most powerful shamans is often referred to as "the language of the birds." A keen attunement to the vocal discourse of the feathered folk has been a necessary survival skill for almost every indigenous community—especially for the active hunters within the group, and for the intermediaries (the magicians or medicine persons) who tend the porous boundary between the human and more-than-human worlds.

There are, however, a few perching birds whose vocalizations are notoriously unreliable—wingeds who wield their words in highly unpredictable ways that seem to vary according to their whim. These are the same species often viewed as tricksters in oral traditions around the world. In many traditions, for example, the raven is rumored to be the most mischievous entity among the birds. For the peoples native to the Pacific Northwest of North America, Raven is generally held to be an outrageously prodigious trickster—far too clever for his own good—an audacious and somewhat self-absorbed magician whose antics actually created (largely by accident) the very imperfect world we now inhabit. In other regions of the world, a primary winged trickster may be Crow, or Magpie, or one of the insatiably curious jays. These birds—ravens, crows, jays, and magpies—compose the main members of the corvid family (Latin: *Corvidae*). The large, long-beaked corvids are all, by and large, omnivores—a trait that they share with human beings, bears, raccoons, and a clutch of other creatures, most of whom exhibit an intense curiosity in their cognitive as well as their gustatory habits. Uncommonly social, long-lived, and weirdly resourceful, corvids are able to adapt themselves to human settlements more brazenly than other birds. Their broad range of utterances seem to be spoken for different reasons in different circumstances and moods; hence their reputation, among some humans, as disreputable scoundrels whose discourse is not to be trusted.

It's a reputation well worth heeding when eavesdropping on birds to glean a sense of what's afoot in the wider landscape. Still, alliances can sometimes be forged with particular corvids. In the Alaskan interior, a collaboration between ravens and hunting wolf packs has been well documented. Soaring above the boreal forest, able to oversee much of what's happening on the ground, ravens will sometimes lead wolves directly to an ailing deer or caribou. Their help is well repaid, since after the kill the wolves allow the ravens to partake of some remains. Human hunters, too, have been known to benefit from the raven's scouting. When my Alaskan colleague, the ethnobiologist Richard Nelson, is out hunting deer to feed his family, he takes particular notice if a raven winging overhead starts to somersault in the sky—as though it wished to capture his attention. He's learned that if he follows that raven's trajectory he'll likely be led straight to his prey, and so will have a chance to reciprocate the favor of the dark-winged trickster.

The feathered ones, then, have long been crucial allies for our kind. Watching them swoop and glide and carve their way through the air surely ignited many of our human aspirations toward freedom and flight. Their bewildering array of colors and chromatic patterns probably provoked many of our earliest acts of self-adornment, while their feathers figured prominently in human rituals and dances frequently influenced by avian courtship displays. Birds have ceaselessly inspired us with their mellifluent voices and polyphonic exchanges, undoubtedly instilling some of our earliest impulses toward song and spoken language. And during the enormously long course of our history as foraging primates, they also aided our survival in a far more immediate and practical manner. For it's they whose cries often alerted us to the approach of dangerous predators. Further, by providing a cover for our stealth that was instinctively trusted throughout the forest, it was often the birds who enabled our best hunters to successfully approach and secure the prey that we needed to eat. They have been for us messengers, intermediaries, envoys from the forest and its wider life, bearers of intelligence we could not do without.

This ancestral sense of the wingeds as messengers, and as guardians of a sort, is preserved for many persons in the conception of angels. (The word "angel" itself comes from the Greek "angelos," meaning "messenger.") The iconography of angels has always shown them with feathered wings. Many contemporary mystics would have us view the wings of angels purely as a visual metaphor, as a way of imagining subtle energies that simply have no physical form. Yet it is likely that the wondrous qualities ascribed to angels were once associated with the elusive, winged presence of birds themselves.

Numerous hunting and gathering cultures have honored birds as emissaries from a more expansive field of intelligence. But with the rise of sedentary civilization and its written-down scriptures, intelligence—as we have seen—was gradually banished from the surrounding world and sequestered within our single species. It became more and more difficult to acknowledge or even recognize other animals as bearers of insight or sources of wisdom. If these winged singers still seemed to grant us a kind of grace even in our settled towns and cities—if there was something about their sudden, swooping visitations that still carried intimations of other dimensions, and something in their songs that touched a forgotten chord within our chest, a fleeting memory of contact with a wider, more ubiquitous awareness—well, then the uncanny nature of these encounters could hardly be attributed to the birds themselves. Since humans were now the sole carriers of consciousness in the earthly world, the unexpected meeting with a small, exceedingly awake feathered presence could only be interpreted as a visitation from a higher, more rarefied kind of human: a little person with wings. Through a concatenation of traditions including Judaism, Zoroastrianism, Christianity, and Neoplatonism, such winged persons came to be viewed as messengers, envoys from a celestial divinity—ultimately as members of the heavenly host, the supernal attendants of God.

Nonetheless, the most common characteristic in the many descriptions of angels, apart from their feathered wings, is their musical nature. *Angels sing.* Indeed, a "choir of angels" regularly heralds the arrival of the Holy One with hosannas of praise, the singing of angels announcing such moments when the divine presence is about to manifest itself, or about to withdraw. This pattern closely parallels the swelling music just before dawn, at birdlight, when the blazing sun is about to enter the great hall of sky, and then later: the rising chorus of song as that same radiance is about to sink below the horizon, and so to depart from the celestial hall.

Is it not obvious, then, that long before angels were conceived as invisible heralds of an unseen deity, they were inseparable from these feathered harbingers of the radiant sun itself? From the winged choir offering up its hosannas of praise every dawn and every dusk? Can we doubt that for our primate ancestors, it was these winged singers, the birds themselves, who were felt to be intermediaries between our ground-bound world and that celestial resplendence, that source, the great god of the day-lit world—whose face, even today, we dare not look upon directly?

After a series of lectures in New England, I wander along the rocky crest of some low mountains above the Maine coast. As I come into a wooded cirque, the song of a hermit thrush rings down from the upper branches. Soon another hermit thrush sounds nearby, and then two more from some slight distance, their several songs overlapping in the needled air, each with a different inflection, yet all of them utterly ethereal. I choose a single singer upon which to focus my listening, and let myself be carried by the song. The first sustained note of each phrase is held for a long moment, followed by three spiraling, rippling turns fading slightly at the end, until another sustained tone launches the next phrase on a different pitch from the last: "Sheeeeee fridiila-fridila-fridila . . . Sooohhh ridileee-ridileee-ridileee . . . Saaaayyy teedelaa-teedelaa-teedelaa . . . Seeeeeeee tidleee-tidleee-tidleee . . ." And it's as though each phrase unlocks a different vertebra along my spine, spilling a ray of light into that node, and then the next phrase unlocks another node—"Sheeeeee teedelaa-teedelaa-teedela . . . Suuuuu ridileee-ridilee-ridilee"—until all my vertebrae have become transparent crystals trading gleams among themselves. The song drops to a middle register, and then a higher vertebra, and then a glimmering tone so high it's almost beyond the reach of my hearing, and as it spirals there I feel the top of my skull opening as I seem to ascend, a single photon, into the sky.

The songs of birds, utterances at the origin of human language, release us from the bounds of our own speech—as their winged forms, watched intently, sometimes release us from the grip of the ground. Birds have long been our chaperones to the heights, as well as emissaries from there, rays of the sun taken solid form.

SLEIGHT-OF-HAND

(*Magic I*)

Mountains slice into the sky, carving the blue into a monster with eight or nine gaping mouths. As I move along the trail, those mouths variously open and close in slow motion, biting back at the sharp-edged peaks. The Himalayas are youthful mountains, only recently shoved skyward by the gradual collision of tectonic plates, their jagged edges not yet smoothed by the gradual erosion of wind and water. The narrow trail I'm following angles down along the slope of a broad ridge to the Sherpa village in the valley.

I am in my middle twenties, trekking through this gargantuan landscape with a young translator, Temba Sherpa. When we first arrived in this village some days ago, huffing and puffing after mounting a long series of switchbacks zigzagging up from the lower valleys, I'd made inquiries regarding a particularly powerful *jhankri,* or medicine person, rumored to live somewhere in the vicinity of this settlement. For that's what drew me up into these peaks: I was traveling as an itinerant sleight-of-hand magician, hoping to learn from the traditional magic practitioners who ply their craft in these regions. In response to my queries regarding the jhankri's whereabouts, we'd been pointed to one of the village houses where the magician's parents lived; they, presumably, would be able to direct us to his dwelling.

Directions, however, were not so easy to obtain. Upon asking his gray-haired parents how we might find their son's home, we were invited to stay with them for the night. I had by then grown accustomed to the curious obliqueness and indirection whereby traditional peoples speak of their magicians; in Indonesia, half a year earlier, I'd been forced to acquire a new depth of patience when trying to make contact with those *dukuns,* or sorcerers, about whom the most powerful stories circulated. I had finally learned to settle into some village in the general terrain of a specific dukun, and to make evident my intent simply by performing an innocuous bit of magic, preferably without calling overmuch attention to myself. When I was walking past the rice paddies in the morning, aware that a few of the farmers—with their ankles and forearms submerged in the water—would likely be looking curiously at me, I might sneeze a few times before reaching my open hand into the space above me and plucking a large, colored handkerchief out of the air. I'd blow my nose in the handkerchief and then reach up to let it vanish back into the space above my head—all with the utmost nonchalance, without interrupting my steps or turning to glance at the farmers. I merely walked on by. And then I would just wait, biding my time in the village, confident that word of this odd happenstance would make its way to the appropriate ears, and that at some moment I'd be contacted by the very magician I was hoping to meet. The contact, when it came a day or two later, was always indirect, most often through a young child who would tug on my sarong while I was reading, or pinch my arm while I was dozing, and then urge me to follow him up through the rice paddies—balancing on the low dikes that separate them—and into the woods.

But this was my first attempt to make contact with a magician here in the mountains of Nepal, and so, not yet familiar with the local etiquette, I decided to try a more direct approach. We accepted the invitation to stay at the home of the shaman's parents, sleeping under yak skins on the second floor with the rest of the family (the first floor, partly dug into the steep hillside, was for the livestock: several yaks and a water buffalo). I was hoping their son, alerted to our visit, might turn up the next morning. If not, I assumed that the couple would give us directions to the jhankri's dwelling that following day. Yet neither of these hopes panned out that next day, or the day after that. Our stay with the elder couple did provide me a chance to acclimatize to an elevation far higher than any I'd yet experienced, wandering the village and quietly adjusting myself to the rhythms of life in a Sherpa household. Like many of the homes in which I would find myself during the next six months, the house itself was elegantly simple, wooden beams and window frames set in a solid structure fashioned of hewn stones, plastered over with clay. Apart from the small partitioned-off room where Temba and I slept, the living quarters were a single long room. Along the front wall of this room was an open fireplace—the stove around which most human interaction seemed to unfold. It was here by the fire that people would gather to talk and to pick potatoes, their skins blackened, out of the hot coals.

The only person who did not come to the fire was an ancient old man, presumably the jhankri's grandfather, sitting always off in the corner of that long room, his head somewhat bowed, his body motionless except for the thumb of his right hand, which was shoving one bead after another along the loop of prayer beads that dangled from his half-open fist. In concert with the flicking beads, a breathy prayer issued like an incessant mutter from his lips: "Om mani padme hung, om mani padme hung, om mani padme hung, om mani padme hung . . ." It was the mantra of Chenrezig, the bodhisattva of compassion—"Ah, the Jewel in the Lotus!"—transmuted by the old man into a kind of breathing. His lips never ceased shaping themselves to this mumbled prayer regardless of whether he was inhaling or exhaling. Yet apart from the minute movements of his lips and that single digit, his bent body was entirely immobile. No one had introduced us to this aged presence when we were taken into the house that first evening, and when I came to sit by the fire the next morning he was in the same corner, in the same stooped position. No one ever addressed him—or even acknowledged him. Although I at first found his muttering presence somewhat unnerving, after a couple of days interacting with the others by the warm fire at mealtime, learning some simple Sherpa phrases, and drinking chang (which seemed a kind of alcoholic yogurt) with Temba and the jhankri's father in the cold evenings, I stopped noticing the old mutterer as anything other than a somewhat broken piece of furniture. Dementia, I noted to myself, is found in every culture.

On our third evening in this house, the father finally gave Temba instructions to the jhankri's cottage, along with a hand-drawn map. The jhankri's home, it now seemed, was many hours' journey from the village; we resolved, nonetheless, to make our way there the following day. I slept well that night, my first decent sleep at this extreme altitude.

When I rose the next morning and came alone into the main room, the fire was still blazing. The jhankri's parents must recently have headed out, and Temba as well. But when I sat down by the flames to forage among the coals, the quality of silence in the room seemed oddly different. Only then did I realize that the wizened old man was missing. He was not in his corner, nor anywhere else for that matter. Curious. Perhaps the others had carried him down the stairs so they could wash him in the open air? Gingerly nudging an ashen potato to the side of the hearth, I took it up in my fingers, tossing it from one hand to the other, back and forth. When the potato was cool enough to hold, I began peeling the black skin as I leaned over to the nearest window, peering out through the smoky glass into the fields. And straightaway dropped the potato. For there, on his own, was the rickety old man standing erect by a wooden fence at the near edge of a field. In his right hand was the looped string of prayer beads, slowly moving through his fingers; I could see his lips slightly opening and closing as the endless prayer rolled out of his mouth. I could not really hear it, although that muttered mantra had been so incessant these last few days that I couldn't help but imagine it as I watched his lips. A raven swerved down to perch on one of the posts about twenty yards down the fence from the old man and his vacant stare. There was no one else to be seen. I had not thought that the old fellow could move his limbs, much less balance upright on his own.

As if to confirm my thought, the old man now appeared to be tipping over. His erect body had lost its precarious balance and was tilting further and further forward as I stared. The man was toppling in slow motion. I thought to cross the room and rush down the stairs, but couldn't take my eyes off him as he tumbled over—his body finally bending at the waist as his right arm, extending downward, broke the fall. And there he now was, pitched forward, with only his feet and one hand touching the ground, his bent body teetering on its right arm. Then the left arm swung stiffly forward to plant its hand beside the right. Good. A stable position at last. I was about to head down to help when I saw him walking his hands slowly back toward his legs until he was able, remarkably, to raise himself to a standing position. It was an amazing recovery. As his arms gradually came back up, I noticed that the string of beads was dangling now from his left fist, rather than his right. I peered hard through the glass to see the right hand, and finally glimpsed it, now clenched, on his other side. I winced. He must've hurt that hand when it broke his slow-motion topple, and so passed the loop of beads to the other fingers.

But I was wrong. He wasn't clenching his right hand in pain. It was clenched because it was holding something. I wouldn't have realized this had not the old man's torso suddenly erupted into fluid motion, the right arm arcing back and around in an outrageously swift movement as it flung—actually whipped—a stone straight down the fence line toward the raven. My wide eyes saw the throw's arrow-straight perfection in the split second before it made contact with feather and bone—but when did it dawn on me that the old man's absurd slow-motion tumble and ever-so-slow rise, moments earlier, had simply been a covert way of picking up a stone without startling his chosen target away from its perch? It could only have been after, for now there was no chance to think before the stone plowed straight into its destination. Except that the raven, in the feather-tipped instant before contact, hopped three inches straight up into the air; the stone whizzed through the vacated space in the single moment before the bird dropped back down to inhabit that space once again, utterly unruffled. The stone sped on past and buried itself in the upturned soil of someone's potato field with a small burst of dust. I stared at the raven, coolly poised on its post as though nothing at all had happened. My mouth agape, I turned my stare back toward the old man: there he was, gazing into space, immobile except for the string of beads slipping one by one through his right fist, and the breathy prayer moving through his lips.

It occurred to me, then, that this man and that raven were probably not strangers to each other—that they had likely been testing one another's skills for a long time.

The close alliance between traditional, indigenous magic practitioners and their animal familiars has by now been reported so much in the modern West that it has become a cliché. As with all such stereotypes, the familiar phrases and hyped images—the witch's black cat, the *brujo*'s animal ally, the medicine person's totem—function to hide (and sometimes to protect) a many-faceted reality whose complexity and practicality remain unexplored, unsuspected even by those today who most have need of its medicine.

In truth, it may not be possible to *write* of the enigmatic interchange between indigenous magicians and their other-than-human counterparts without falsifying this interchange. For as I've indicated over and again, the written word regularly draws both reader and writer into a style of awareness that is mostly closed to other species. Our commonplace practice of reading texts has its deepest ancestral source in the indigenous hunter's careful reading of animal tracks pressed into the surface of the earth. But the recent transformation of this ancient craft into a purely human reading of our *own* tracks, trailing off across the page or screen, has made possible a degree of human self-involvement entirely unknown to our oral ancestors. Today, printed words such as these you're reading circumscribe a symbolic space of communication that we carry on solely among ourselves, a hyperreflective style of cogitation far too abstract to register the incarnate intelligence of other animals (often so detached that it renders us, while we're reading, oddly impervious to the promptings of our own animal flesh). The powerful, self-enclosed spell of the written letters easily eclipses that subtler magic—the nuanced exchange between the human animal and the animate earth.

How then to write of an unexpected dialogue that once happened between me and a shining spider, or the vital intelligence that confronted me once in the person of an immense grizzly after my kayak ran aground in a stream thick with spawning salmon? How to write of the moment, years ago, when while pondering a patch of crinkled lichen on a protrusion of granite I suddenly found myself falling down and down into the depths of that lichen, tumbling head over heels into a limitless and baroque universe of folded crimson?

And how to write of the odd sensations that came over me this day, as Temba and I hiked up the valley toward the home of the jhankri? I had noticed, months earlier in Indonesia, that my body would sometimes feel peculiar upon coming into the presence of a particularly powerful sorcerer—a feeling that would pass after a few days. I had at first misconstrued the experience to be that of some illness, a vague flu that I happened to come down with at that unfortunate moment. But after contracting a similar illness when I encountered another unusually strong dukun, I began to realize that it was not an illness at all—that I was misinterpreting sensations that simply were very new to my organism. I couldn't explain this odd feeling within my limbs, yet once I realized that it wasn't noxious, I found that there was nothing inherently unpleasant in the feeling; it was, simply, strange. And so I had learned to recognize the impressions, when they arrived, as an inward indication that I was standing close to a magician of unusual depth and strength.

Now, however, I was merely on the way toward the home of a healer I had not yet met—in truth I was still a half day away—yet already I could feel that same tingling along my skin and within the muscles of my arms and legs. Not that the feeling was precisely identical to that which I'd experienced those several times in Indonesia, but when I began wondering if I was getting sick, I abruptly realized that the sensations had the same telltale quality as those I'd felt before. Could it be that my organism was already sensing the jhankri's presence? Or was my body merely anticipating the imminent encounter with this curious person, of whom I had heard only rumors?

In any case, I soon became aware of a new quality in this experience: the throb in my muscles was such that I could no longer wholly straighten my legs—or rather, it felt uncomfortable to fully straighten them. So while hiking along the trail I was no longer entirely extending each leg after its foot came down, as is our customary way of walking; instead my legs stayed somewhat bent even at their point of greatest extension. And as a result my stride now took on a springiness, a bounce, that was very new to me. You might think that such a way of walking would be tiring, especially when carrying a heavy pack, but in truth I found it increasingly pleasureful, a new and more playful relation to gravity. I noticed that my arms were oddly relaxed as well, dangling loose and swinging in the gravitational field. It seemed as though I was conscious of this enveloping field of force for the first time in my life, and could feel myself dancing within it, exploring the quality of gravity as though it were an elastic fabric stretched around me. This lucid sense of space was accentuated by the clarity of the many smells drifting through my awareness, my nostrils flaring at the scents rising from the dirt and from the various herbs we brushed past as we hiked. My nose, somehow, had come strangely awake.

This weirdly limber and prescient state stayed with me for the rest of our mostly uphill hike, which at long last brought us to the home of the jhankri. The modest stone house was set among a clutch of large boulders, close by the edge of a broad and very deep canyon. A narrow river wound its way far below. When we knocked, the door was opened by a woman with small, shining eyes, wearing a dark robe with a multihued square of cloth hanging from her waist. We bowed a simple greeting, after which Temba, as my translator, seemed to get tongue-tied and flustered. But she ushered us in. I set my backpack by the door; Temba did the same. Our hostess placed a pot of water, for tea, onto the fire.

My eyes had trouble adjusting to the dark; I could see only that the woman, by the light of the fire, was gesturing me toward the far end of the single long room. Still mostly blind, I took a few steps in that direction and stopped, waiting for the darkness to dissipate. I gradually became aware of the flickering glow on the high cheekbones and forehead of an upturned face. I placed my hands together and bowed gently—"Namasté"—but there was no reply. The firelit face was low to the ground, turning slowly from side to side; abruptly I realized that the turning head was sniffing the air on one side and then the other, scroonching its nose rather like an ape as it did so, widening its nostrils to better sample the air as I approached. The jhankri was squatting in a balanced crouch, forearms resting on knees. I moved to the side wall and settled onto the floor. He continued sniffing the air with short double inhalations, his upper lip rising and falling, and I felt the hairs rising on the back of my neck. Here I was in the close presence of an animal—a large primate so intensely aware and wary of my presence that my own animal hackles were aroused!

The jhankri's wife approached our end of the room and set down two cups of steaming yak-butter tea on a low wooden bench. I looked around for Temba; he was still standing by the door where we'd come in. I called him to come sit down. But he didn't budge. The jhankri stared at me, then abruptly raised himself and walked over toward the door with a strange, gangly grace. I saw that his knees never fully straightened; his legs stayed bent like taut springs as he moved through the thick air. I recognized the odd gait that my body had taken up that afternoon—noticed its primate-like resonance as the jhankri spoke a few words to Temba and led him back to sit with us—and I felt myself sliding further into a trance. My eyes, honed in some way by the singular state that had taken hold of me, could now see the currents of air whirling in small vortices and eddies as the jhankri's limbs sliced through the space . . .

Or were those eddies now visible by virtue of some residual smoke that hadn't made it out the smoke-hole in the roof?

As the jhankri sat back down, I could sense the floor under my buttocks, and the bones within my buttocks. I could feel my skull as well, and the segmented tower of the spine holding my skull aloft, suspended in space. When I shifted position my spine swayed slightly, like the limber trunk of a young sapling bending in the breeze, until it found a new poise with my head balanced atop it.

Somehow, in this curious house, I could sense my materiality much more vividly, the physical density of the several layers of my body—and so could feel as well the solidarity between my thingly presence and various plants, even my kinship with the stones in the wall and the boulders outside the cottage.

The jhankri was staring at my face, sweeping his gaze across my skin and the various tracks or traces to be found there. I asked Temba to translate for me as I thanked the householders for the yak-butter tea. But Temba shook his head. So I offered my thanks directly to the jhankri, whereupon he spoke some words to me, gesturing toward the tea. When I raised my eyebrows toward my translator, hoping for some clarification, Temba simply shook his head and whispered, under his breath, "No . . . No. I cannot translate. We go now." I asked him if the jhankri had told us to leave. "No. I cannot translate for you. We must go."

I saw that Temba would not look directly at the jhankri. Yet our host, smiling slightly, was pushing the teacups toward us across the bench. I mumbled to Temba that surely we should drink the tea before leaving. But his hands were shaking, and one of them was tugging on my jacket: "We must go now. We go." His anguish was palpable. "Now?" I asked. "Yes." The tremble in his arms was shaking me out of the trance. Thoroughly flummoxed by the situation, I climbed to my feet and bowed again to the man, trying to thank him and his wife and to apologize for leaving so quickly, but Temba would not translate any of this—he simply pushed me toward the door. I reached for my backpack and dragged it outside. The door closed.

I waited until we'd stepped past one of the boulders, then turned to Temba: "What's going on?"

"It is bad. Very bad," he said. "I should not have brought you here. It will be bad for my family. They probably get sick. This jhankri may make them sick. Maybe I will die."

"Did the jhankri say that he would make your family sick?"

"I could feel it when we went in the house—"

"But did the jhankri say that he would hurt your family? Did he tell you it was bad that we came?"

"I don't know. But I should not have brought you here. It is a terrible thing." He turned and started walking down the trail. It was late dusk, and I had no idea where we were headed. I leaned over my pack and started digging for my flashlight. Yet when I realized that Temba wasn't waiting, I hoisted the pack and followed him down the darkening path.

We slept that night above a rushing stream, our sleeping bags huddled against each other for warmth. In the morning we decided to head toward the village where his family lived, and where he'd grown up. Temba had only recently moved to the big city, Kathmandu, in hopes of securing work as a mountain guide. I had been introduced to Temba some weeks earlier at the Sherpa Cooperative in Kathmandu, in a last-ditch effort by one of the directors to locate a translator willing to accompany me on my quest to learn from traditional jhankris in the high mountains. I had already been paired up, twice, with two different guides; each had backed out of the trek upon discovering that I was entirely serious in my intent to meet with the mountain shamans. Temba was the third and by far the youngest of those who signed on to the venture. This time I had been as explicit as I could, spelling out my aims at length while we were still standing in the office of the Sherpa Cooperative. Temba had insisted, then, that he had no problem with my interest in traditional magic, and that he'd have no difficulty translating for me in any interactions with the Sherpa jhankris. And so I had hired him as my guide and translator.

Yet now, in our very first encounter with a shaman from his own culture, Temba had been stricken with fear, almost paralyzed, hardly able to breathe in the jhankri's presence, much less translate for me.

Already in Indonesia I'd become familiar with the fear that some dukuns seemed to inspire, even among the very villagers whom they helped back to health. It was often assumed that a shaman, in order to prevail against the harmful spirits that can sap the strength from a grown person, must be well acquainted with such spirits, and may even be consorting with such demons in the off-hours. This was especially the case with regard to the most powerful magicians, those who had the greatest success in curing diseases, easing community disruptions, or even altering problematic weather patterns. If such a shaman could bargain so successfully with the malevolent energies, regularly convincing them to back off, it seemed likely that he or she was on companionable terms with those shadowed powers.

Learning closely from several magicians in the Indonesian archipelago had exposed me to the profusion of rumors steadily circulating through each island's population regarding the practice of harmful (or "left-handed") magic by certain magicians or sorcerers. I'd slowly come to recognize that the abundance of these suspicions far outweighed the actual extent of such practices. The proliferating suspicions were less a consequence of actual acts than of the most common diagnosis offered by medicine persons to explain the ailments they work to heal. A diagnosis of *witchcraft*—a suggestion that a client's illness was consciously inflicted by the deployment of *bad* (or malevolent) magic—readily implies that the disease can be alleviated by the diligent application of *good* magic. Indeed, such a diagnosis will often hasten the curative efficacy of such good magic or medicine. The resulting abundance of dark diagnoses, however, inevitably means that most shamans and medicine persons—most of those who work with magic—will occupy a very ambivalent place in their communities.

I had at first been surprised that the several shamans I came to know in Indonesia did little to counter the many suspicions and rumors that swirled around them. Yet I soon came to discern that the aura of danger surrounding the most potent magicians had an important benefit: it served to protect those practitioners from what would otherwise have been a ceaseless onslaught of requests for ritual assistance. The rumors ensured that only those persons who were in dire need of help would dare seek out the magician and his or her powerful craft.

Temba was young. I should not have been surprised by his reaction to the jhankri, given the response of the earlier translators back in Kathmandu. It was clear to me, from the effects I'd felt in my body when at his home, that this jhankri's craft was at least as intense as any I'd encountered in Indonesia, and I was terribly frustrated that our meeting had been cut short. Whatever else I'd experienced, I'd felt no malevolence in his presence. Nonetheless, Temba's still-evident dismay—and his fear for his own family—prompted some alarm. I was concerned for my teenage guide, and my sense of responsibility in all this began to thicken as the day went on. I was eager to get Temba back to his village, and to ascertain that his folks were all right.

I resolved to return to the jhankri's home on my own at a later time.

By late afternoon we reached the monastery where Temba's younger brother lived as a monk, and made arrangements to stay there for the night. That evening, I asked both brothers to begin tutoring me in the basics of the Sherpa language, and we stayed up much of the night trading terms and phrases. The next afternoon Temba and I finally made it to the small village where he'd grown up, and were welcomed into his parents' house. Temba was hugely relieved to see they were in good health, and they were relieved to have some help digging up their potato fields. I moved into their home (upstairs, again, above the yaks' quarters) for over a week, making steady progress on the language with Temba, trying out my clumsy skills on the neighbors while working in the fields. Finally, confident enough of my ability to communicate simple matters, I took my leave of the household—paying Temba well for the brief time we'd been together. I had originally thought I'd have his services for several months. I felt lousy backing out of our contract; it was now clear, however, that I would never learn from any of the Himalayan jhankris unless I approached them on my own. Temba stayed in his village; he'd help his folks for a time and then trek back down to the Kathmandu valley in search of a climbing expedition to sign on to, or any other job less nerve-racking than this one he'd mistakenly taken on.

I was alone with the great mountains now, and with the god-haunted winds, clambering over rivers on wobbling, ramshackle bridges and following foot trails winding above villages where pungent smoke rose from different hearthfires like prayers dissipating in the morning air. Flocks of small birds swerved and dipped in perfect unison, alighting on a roof and then all rising a moment later to undulate and billow in waves like a windblown cloth before settling in the branches of clustered trees. On some mornings the ringing of prayer bells sounded far-off; on others I remember the loud slap-slapping of prayer flags strung overhead as the breeze ingested the prayers printed on the colored fabric. Now the wind would carry those blessings wherever it blew . . .

On one afternoon, as I was hiking along a small river, a noisy clattering assailed my ears. Rounding a bend, I discovered the racket emanating from a wooden prayer wheel spinning above the middle of the current, a brightly painted cylinder whirling its carved prayers 'round and 'round, propelled by a simple waterwheel at its base.

Wherever I found myself at dusk, I'd make my way toward the nearest apparent settlement, where I was always offered food and a warm corner to sleep, then sent off in the morning with clear directions for my onward journey.

One morning I followed the dissonance of some blaring horns to a small semicircle of lamas and monks chanting in full regalia on a mountainside. They saw me watching from down the slope, and waved me over with big grins. I sat and joined them for the afternoon around their fire, charmed by the cacophony of the horns and by the throaty, bass voices intoning the liturgy, and the sweetness of the rice cake offerings I was urged to eat, and the many birds soaring overhead, apparently drawn by the festivities. Only after several hours, when I stepped upslope to urinate behind some trees, did I discover, with a shock, that upon the rock platform just beside where I'd been sitting (just above my seated line of sight) was the naked, disemboweled corpse of a young woman! I felt the blood drain from my face. Unfortunately I was already walking back down to my spot when I first saw that splayed, bird-pecked body, its innards still attached yet laid open on the rock, and so could not easily slink away. I tried to contain my nausea as I dizzily took my seat at what I now realized was a funerary rite. The festive celebration I'd been enjoying for several hours abruptly morphed into a macabre, haunted ritual, stained by my own fear. The clamor of the horns now seemed sinister to my ears, and strong smells that I'd thought were from the incense now assailed my nostrils as a threat. Even the blue sky itself began to seem an oppressive power as it imbibed the dark smoke of the central fire.

I had never seen a corpse displayed in such a manner, much less sat beside one for a whole afternoon. As I slowly gathered my rattled wits, I came to realize that the deceased woman's body was being offered to the animal powers and the spirits of the air. Hence the broad-winged lammergeiers and the other carrion birds that'd been circling overhead all afternoon. Death was not hidden here as it was where I came from, not cosmeticized to mask its unnerving power, nor embalmed to stave off decay. Instead it was displayed in all its matter-of-fact awfulness as the ever-present underside of existence; indeed, here it was being offered as food to the wilder life that surrounds us. I was stunned, grappling with the demons stirred inside me by what was happening. Yet the monks were smiling and sometimes laughing, at ease with their prayers as they guided the woman's spirit through the *bardo*—through the gap between worlds. By the time I left the gathering, bowing low to the various lamas, my idyllic sense of these mountains, and the prayer-infused breezes that move between them, had deepened considerably to make space for the earthen tones of danger, disease, and death that color life in this stark terrain. To make room in my awareness for the loss and racking grief that shadows life on this planet, the rain of tears that somehow enables all this staggering beauty. Only after this encounter was I able, for the first time, to return the ferocious gaze of Mahakala—the horrifying, wrathful form of the Buddha—who glared down from the inner walls of the small monasteries that I came upon in these high valleys.

At one such gompa, the resident lama was startled into laughter by my sleight-of-hand metamorphosis of a simple, weathered stone into another, carved with prayers, that I'd found some days earlier. He took my hand and led me down a long trail to the river, so I could watch his two students as they worked with temple woodblocks artfully carved with Tibetan ritual verses. Normally these precious woodblocks were used to print out liturgical books. But now—amazingly!—the students were stamping the woodblocks over and over into the flowing surface of the river, so that the water would carry those printed prayers to the many lands through which it traveled on its long way to the Indian Ocean.

Here, remarkably, was a culture wherein written letters were not used merely as a record of words once spoken, or as a score for oral speech, but as efficacious forces in their own right. The letters were not just passive signs, but energetic agents actively affecting the space around them. Whether written on the page of a book or carved into woodblocks, whether etched into standing stones or printed on flags, the Tibetan letters held a power that could be activated not only by human beings but by insects crawling through their cracks, and by water flowing along their shapes, and even by the breeze gusting across them. Human intentions, carried in dreams and prayers, mingled here with the intentions of stones, trees, and rivers. Clearly, "mind" in this mountain region was not a human possession; it was a power proper to every part of the elemental field.

We participate in this encompassing awareness with the whole of our body, as other animals participate in it with theirs, the snow leopard with its tensed muscles and the hawk with its splayed wing feathers. Every creature here inhabits and moves through the same field of mountains and melting ice, imbibing the same air, the same boulder-strewn awareness. Yet each animal filters this awareness with its particular senses, its access to the whole limited by the arrangement of its limbs and the specific style of its pleasures, by the way it obtains nourishment and the way it avoids becoming food for others. Each creature—two-leggeds included—has only a restricted access to the mystery of the real. As a human I may have compiled a great mass of data about the ways of the world, yet in a practical, visceral sense (carnal knowledge being the primary form of intelligence), an earthworm knows far more about the life of the soil than I do, as a swallow knows more about the wind. To be human is to have a very limited access to what is.

Science has tried to push past the carnal constraints on our knowledge by joining deductive reason to the judicious application of experiment. Traditional, tribal magicians or medicine persons take a different approach. They seek to augment the limitations of their specifically human senses by binding their attention to the ways of another animal. Steadily training his focus upon the patterned behavior of another creature—observing it closely in its own terrain, following and interpreting its tracks, becoming familiar with its calls and its styles of stalking or foraging—the medicine person renders himself vulnerable to another, non-human form of experience. The more studiously an apprentice magician watches the other creature from a stance of humility, learning to mimic its cries and to dance its various movements, the more thoroughly his nervous system is joined to another set of senses—thereby gaining a kind of stereoscopic access to the world, a keener perception of the biosphere's manifold depth and dimensionality.

Like anything focused upon so intently, the animal ally will begin visiting the novice shaman's dreams, imparting understandings wholly inaccessible to her waking mind. She may spend a whole night journeying as that other animal, stalking her prey and sometimes killing and devouring it, before awakening in this two-legged form. Most importantly, because the young shaman is now informed by two very different sets of senses, her allegiance to her own single species begins to loosen; she begins to catch glimpses of a shimmering, ever-shifting lattice of affiliations and interdependencies—the filamental web that binds all beings. Now and then she may catch herself pondering matters less from a human angle than from the perspective of the forest or the river valley as a whole . . .

In the course of these first months in the Himalayas I came into contact with several jhankris of very diverse skills, and I lived for several weeks with two of them, a husband and wife who were both highly regarded as healers. The woman was suspicious of me at first, since rumors of my own craft had preceded me up their valley. I'd made the mistake of doing some sleight-of-hand for a ragamuffin band of children in a small village, changing stones into Nepali coins and then back again before making them vanish into the air. News of these simple feats had spread quickly among the neighboring villages, arousing both curiosity and some alarm among the adult Sherpas. While magic of various forms had a clear place in this culture, it was odd for a white Westerner to even *believe* in magic, much less to display any such rapport with the invisibles. So the female jhankri glared at me when I first made my way to their hut; I realized that I was not welcome. Only by returning day after day, offering to help carry water up from the rushing stream, was I able to ease the initial antagonism. When her husband finally challenged me to show something of my own skill, I produced a brightly colored square of Sherpa cloth out of the air, then formed a small, empty pocket in the middle of the cloth, letting it dangle down from my fist. I invited the jhankri to blow upon the fabric. Then, singing softly, I opened the folds to reveal a gleaming quartz crystal. (The crystal was a talisman of mine that I'd been carrying in my pocket for most of a year.) The shaman took the quartz in his hand, holding it up to the sun. Then closing one eye, he brought the crystal close to the other, peering at me through its facets as he turned it this way and that. He asked if there were other crystals wanting to appear. I told him that I did not know, but that I hoped he would accept that crystal as a gift from the spirits.

Perhaps I could teach him some of the mantras that I used for my magic? Of course I would, yes, if in exchange he would allow me to sit in on some of his healing sessions . . .

It was a deal.

I moved into their household from the smaller family home where I'd been sleeping, and soon found myself accompanying the man, Dorjee, to a range of healings—usually in the evening and sometimes lasting the whole night.[n] Some of the healings were much stronger, in their feel, than others. In the most intense sessions Dorjee, aided by the rhythmic pounding on his *dhyāngro* (a large two-sided drum), slipped into a kind of delirium.

Meanwhile, during the days, I shared with him some basic techniques central to my practice of legerdemain, painstakingly walking his fingers through the most elementary sleights. However, apart from letting me attend the healings, the jhankri shared little or nothing of his craft with me. There was a younger Sherpa man who sometimes assisted Dorjee; the shaman's comments or instructions were reserved for him. Much of what unfolded during the healing sessions was entirely opaque to my understanding; I could only witness the ritual gestures, letting myself be carried by the rhythms or distracted by the keen concentration in the faces of the others present.

But then, at the frenzied height of one such session, I was startled to see the shaman extract a bloody, tumorlike knot of matter from the side of a feverish woman's abdomen. It was a remarkable happening, and had a powerful effect upon those watching. Only then did I realize that certain sleight-of-hand methods were already a part of Dorjee's toolkit. For my conjuror's eyes had glimpsed that bloody gizzard hidden in Dorjee's palm for several moments before it was extracted from the client's body.

The more I thought about it, the more I wondered whether Dorjee drew upon such legerdemain that evening simply because I was watching, as a way of displaying his own skill at something akin

[n] I have altered the names in this chapter and the next, in order to respect both the privacy and the practice of the individuals mentioned.

to my craft. When we were alone two days later I cautiously inquired about the extraction, praising his skill yet making clear that I had seen the deception involved. The shaman took no offense. Such a technique can be used, he said, only when other approaches will not work. I waited for him to explain. Dorjee sat down on the ground. Here is what he shared, at least as far as I could understand (putting here into my terms what he conveyed not only with words but with many gestures and facial grimaces):

There are certain demons whose grip on a sick person's innards is so firm, so tenacious, that they cannot be dispelled by the jhankri's *phurba* (a sort of magic dagger), or by any other means. No amount of trancework can induce such a demon to loosen its grip. It is in such cases that the jhankri must employ a bit of subterfuge to trick the demon out of the client's body. For that malevolent knot of energy will not depart its host's body unless it is provided with *another* physical body to inhabit. This can be a smallish object, as long as it is suitably sanguine to attract the demon (demons apparently love gruesomeness and gore). The bloody gizzard of a large bird will do nicely.

But here's the rub: The demon must not be allowed to catch sight of this object beforehand, lest the demon become suspicious and realize the trap that is being set for it. Since a demon perceives the world through the senses of whatever body it currently possesses, the jhankri must therefore keep that bloody gizzard entirely hidden from the patient's eyes. Once the patient herself has been lulled into a trance by the jhankri's drumming and dancing, then the demon within her will also be in a trance, and so can be gradually lured up out of the innards toward the body's surface, or skin, by various means. It is then, and only then, that the gory object must come into view, at that precise spot on the patient's skin—as though it were being extracted from her body. For then the unsuspecting demon will be unable to resist being drawn across the skin by the proximity of the delectable, gristly form that has just appeared there. It's as though that bloody gizzard or other object effectively sucks the demonic spirit into itself. The jhankri can now display that bloody object to the patient and the others present, before tossing it into the fire. If the patient recovers from her illness, the jhankri will know that the sleight-of-hand was effective, not in tricking the patient (as a Westerner might claim) but in tricking the demon—the actual illness—itself!

Dorjee was wholly earnest in his conviction that it was the demon who was being deceived, not the client or the client's family. Because I had recognized the subterfuge employed, Dorjee insisted that I accompany him eight or nine days later to the village of the woman who'd received the treatment that evening. She was tilling one of the village fields when we arrived. She affirmed that the long-standing, debilitating pain in her midriff had eased and then gone away in the days following the healing, and had not returned. Her energy, meanwhile, had been slowly and steadily renewing itself; she felt well.

The healing by Dorjee was one of several uses of perceptual legerdemain that I have witnessed by traditional medicine persons. I came to believe that the shamanic application of sleight-of-hand for healing purposes is likely the aboriginal source of the entire craft of sleight-of-hand conjuring. For the skillful practice of sleight-of-hand magic is a kind of shock treatment for the person watching—a way of jarring his nervous system, and immune system, via the direct conduit of his senses. It is a potent technique for disrupting frozen patterns and fears, knocking loose the regenerative capacity of the body. Contrary to modern assumptions, sleight-of-hand conjuring probably originates not as an illusory depiction of *super*natural events, but as a practical technique for unlocking, and activating, the fluid magic of nature itself.

Almost three months had passed before I made it back to the home of the powerful jhankri whom Temba and I originally met on our arrival in the high Himalayas. It had taken that long for me to feel even minimally attuned to the rhythms of life in this massive landscape. I had not wanted to return until I felt reasonably able to communicate clearly, albeit clumsily, by combining a growing repertoire of Sherpa words and phrases with a range of gestures. When Temba and I had made our awkward departure from the house of that jhankri, it was obvious that I'd have to be far better prepared before returning to that isolated place among the clustered boulders. The peculiar circumstances of that initial encounter had never been far from my reflections in the intervening months. I had two dreams during this time wherein I'd been met by that jhankri with his primate gaze. He did not speak during these oneric encounters; in one of them, he simply pointed out with his flaring nostrils some aspect of the ambient landscape to which I had not yet paid much attention. The day after that dream I indeed found myself more lucidly awake to the shifting hue of the icy peaks, and the way the cloud shadows as they moved followed the irregular contours of the mountains.

And now as I trekked along the narrow trail toward the jhankri's home—approaching from the opposite direction than I had come the first time—I again found myself distinguishing the variegated scents drifting past my nostrils. Once again I realized that my knees were reluctant to fully straighten themselves, and found myself bouncing with unusual ease along the slope, my muscles responding with rare grace to the subtlest alterations in the terrain. Was my body drawn into this prescient state by virtue of aiming my steps back toward the dwelling of this particular magician? Or was I merely reactivating body memories from that trek three months earlier?

I do not know. I know only that once I discovered again their rock house camouflaged among the boulders and knocked solidly upon the door—and once the door was opened by the jhankri and I found myself gazing into that face slowly breaking into a grin—I was all but overcome by a sense of familiarity that reverberated through my organism, as though I had never really left this place. It was then, I suppose, that my apprenticeship began.

The strangest thing about my time with Sonam and his wife, Jangmu, was how deeply I came home to myself during those days and nights. Rather than sampling alien practices and exploring beliefs entirely new to me, it was the quality of my own felt experience that became ever more fascinating, the carnal thickness underlying even my most ephemeral daydreams. From that first evening in their house, I found myself noticing ordinary, physical sensations much more vividly than I had realized was possible. As though something in my hosts' way of moving somehow untied and dispersed all my abstract reflections. The churning of words within my head simply fell silent when I was anywhere around Sonam, freeing my awareness to witness the unique intensities of particular textures, smells, and sounds as these registered along my skin or in the depths of my viscera. Their home, with its stone walls, had a palpable density that hunkered close as I slept on the mud-caked floor across from Sonam and Jangmu, and when I awoke in the mornings I seemed to emerge from my private dreams into the wider dreaming of this breathing house nested within the broad imagination of the bouldered hillside. My hosts were already at work, whether feeding their few animals or hauling water back from the stream or consulting the spirits regarding the faltering crop of potatoes in a nearby village. Later I would be carrying fallen deadwood gathered from a stand of trees by the far-below river, walking up the switchback trail behind Jangmu—she seeming to float up the steep trail in her bare feet while her back was bent forward, its huge load slung from a single rope tumpline around her forehead, me straining and staggering in my hiking boots with a far smaller stash of fuel under my arms. I remember how completely those walks annihilated any separation of my conscious thoughts from my aching shoulders and my hammering heartbeat and the step and stumble of my legs.

And herein was the strangeness: the more my consciousness sank into the muscled thickness of my animal flesh, the more I could feel the tangible earth around me swell and breathe and move within itself—trees, riverbanks, and boulders quietly responding to all the happenings in their vicinity. It seemed the ground itself felt my footfalls and nudged my steps in the most serendipitous directions, ensuring that I'd come upon some unexpected event at just the right moment—that I'd encounter a hawk just as it swerved into a tree to feed its nestlings, or that I'd step into the precise spot to glimpse, through a momentary opening in the monsoon clouds, two mountain goats coupling on the high ridge. As though by dissolving my detached cogitations into the sensory curiosity of my body, I had slipped into alignment with the sentience of the land itself. Awakening as this upright, wide-eyed, smooth-skinned thing, I noticed that all the other things around me were also awake . . .

It was as profound an experience of magic as any I'd yet tasted, and yet it was entirely ordinary. There was nothing extraordinary about it, not in the least. It was not the encounter with a supernatural dimension that unfurls somewhere *beyond* my everyday awareness, into which I might elevate myself now and then, but with a dimension always operative *beneath* my conventional consciousness, a carnal realm wherein my animal body was engaged in this ongoing interchange with the animate earth. Hence I began to feel far more palpably present, and real, to the rocks and the shadowed cliffs than I'd felt before. I felt that I was known to these mountains now. This experience—this awareness of my elemental, thingly presence to the tangible things that surround me—has remained, for me, the purest hallmark of magic, the very signature of its uttermost reality. Magic doesn't sweep you away; it gathers you up into the body of the present moment so thoroughly that all your *explanations* fall away: the ordinary, in all its plain and simple outrageousness, begins to shine—to become luminously, impossibly so. Every facet of the world is awake, and you within it.

The deeper I slid into the material density of the real, the more I found that there was nothing determinate or predictable about existence. Actuality, this inexhaustible mystery, cannot be domesticated. It is wildness incarnate. Reality shapeshifts.

Eight days after I arrived in his home, Sonam and I hiked to a village several hours distant, where he'd been summoned by what seemed to me a kind of crazy woman. She had appeared the day before, with one eye half shut and the other overly wide, her matted hair so long unwashed that it stuck out suspended on one side of her head like the fire-blackened branches of a tree. She stared at me and spat on the ground a lot; when Sonam turned up, they spoke briefly before she slipped away. The next morning, he and I walked up a long side valley to reach the village. Apparently there had been two accidental but unrelated deaths in the settlement in recent days, and now a young boy was very ill; the people were alarmed and hoping to forestall any further bad luck. When we arrived Sonam stayed mostly on the outskirts of the village fields, consulting only one or two men that he encountered. He worked his way around the wide perimeter throughout the day, turning over mossy rocks, reassembling several half-tumbled chortens, climbing up into one of the few trees—all the while singing to himself—and at one point (as I noted in my journal) tossing some handfuls of water from an icy stream into the air with a shout. I followed him the whole time, yet at a respectful distance, not really sure what I should be doing and trying, unsuccessfully, to make sense of what he was up to. At one point several children brought us some potato bread still warm from a fire, which I alone ate. I had assumed Sonam would enact some sort of healing when we got to the town, but that was hardly the case. He didn't even enter any of the houses. (In truth, I never saw my host overtly involved in any curing ceremonies, although an elderly Bön hermit who lived up-valley told me how, many years earlier, Sonam had brought back from near death another jhankri who at that time was a major healer in the region. But Sonam, as far as I could tell, did not work as a healer. He kept his distance from other people, and they kept their distance from him—although he was called upon, when necessary, to ease darker problems afflicting whole villages or stretches of forest, or even, as I'd heard, to alleviate the calamitous weather endangering a climbing expedition that had employed many Sherpas from the valley.)

Toward the end of the afternoon, Sonam indicated that he was done, and that we should head home. Could we not, I asked, accept some yak-butter tea in one of the houses before we walked back? No: he did not want to speak with the villagers. He bent down to tie several plants and twigs gathered during his work into a small bundle, which he gave me to place in my daypack. Refusing a last piece of the potato bread I'd kept in my pocket, he swung around and headed up the trail. I shouldered the pack and set out after him. The trail was the same we'd taken early that morning in the other direction, winding upslope and then along a steep incline that at times gave way to a precipitous drop-off to our left. Only when I leaned over the edge did the hiss of the river five hundred feet below separate itself from the sigh of the wind pouring around us. Prone to vertigo, at such places I made my way slowly, concentrating on smoothing my breathing while hewing close to the upslope side of the trail. In this manner I would fall well behind Sonam, who, twice my age, glided across the steep mountainside with a careless ease I could not fathom, his knees bent and his arms loosely outstretched, hands dangling limp like a drunken monkey. Sometimes he'd mock my fear by leaping onto a large rock at one side or the other of the trail, landing atop it with just one foot and then balancing there—the other foot swinging beside him—gazing around seemingly at ease until I finally caught up.

At one point I emerged from a harrowing stretch to see Sonam far off in front of me, and so I hustled along the widened path to catch up. The sun slid behind the mountains and a chill rose from the canyon. I had gotten to within forty or so feet of him when Sonam rounded a fold in the slope and vanished from view. As I approached that bend I heard the guttural squawk of a raven, loud, from somewhere nearby. After a moment I heard it again, and then as I rounded the bend I finally saw the raven, poised atop a boulder jutting out over the gorge, to the left of the trail. The bird was facing across the trail; as I watched, it hopped twice to angle itself more toward me, its eyes blinking like camera shutters as it cocked its head. It uttered another more subdued "squaaark" and then hopped down onto the trail. Yet as it did so the raven suddenly seemed to swerve toward me, for it expanded rapidly in apparent size. My arms instinctively flew up to shield my face, but then the bird simply alighted in the middle of the trail. Still, there was something all wrong about the way the raven landed on the dirt—its shape was contorted somehow, and the landing much too loud, until I realized that I was looking at Sonam, and not a raven at all. I blinked. And realized to my utter perplexity that Sonam was much farther from me than where the bird had been. What the hell was happening? I took a few steps toward him, whereupon my eyes discovered that the boulder on which the raven had perched was itself much farther than I'd thought, not eight feet in front of me but about twenty-eight feet away, and hence was a hell of a lot bigger than I'd perceived it to be. So it was this much larger rock from which the raven had just jumped onto the trail, transforming its feathers into clothes in the process. I stood there in a kind of shock, straining to fathom what I'd just seen. Straining even to *allow* what I'd just seen: a man turned into a raven, and then back again. A man I knew. A perfectly impossible metamorphosis had just unfolded before my blinking eyes.

Sonam was looking at me from where he was, his head tilted to the side. I could feel a tremor riding upward from my feet, my legs starting to buckle. I eased off the backpack in order to set it down, but Sonam walked over and hoisted it on his shoulders, motioning that we'd better keep going, as though nothing unusual had just happened. I took a few steps, then stopped and stared around, shaking my head rapidly, trying to make sure that what I was seeing was really there. Had what I'd just witnessed been merely a mistaken perception, a momentary confusion of my senses . . . ? Sonam strode past me and headed along the trail. After a few yards he spun around and stared hard into my eyes. He swiveled his head from one side to the other, then opened his mouth and squawked—a perfect series of raven's caws—before turning and leading the way down in the blue mountain dusk.

SHAPESHIFTING

(Magic II)

The human body is precisely our capacity for metamorphosis. We mistakenly think of our flesh as a fixed and finite form, a neatly bounded package of muscle and bone and bottled electricity, with blood surging its looping boulevards and byways. But even the most cursory pondering of the body's manifold entanglements—its erotic draw toward other bodies; its incessant negotiation with that grander eros we call "gravity"; its dependence upon cloudbursts not just to quench its thirst but to enliven and fructify the various plants that it plucks, chomps, and swallows; its imbroglio with those very plants and a few animals, drawing nourishment from them for its muscles, skin, and senses before passing that chomped matter back to the world as compost that might, if we were frugal, be used to nourish the soils in which those plants sprout; its bedazzlement by birdsong; its pleasure at throwing stones into water and through glass; its mute seduction by the moon—suffices to make evident that the body is less a self-enclosed sack than a realm wherein the diverse textures and colors of the world meet up with one another. The body is a place where clouds, earthworms, guitars, clucking hens, and clear-cut hillsides all converge, forging alliances, mergers, and metamorphoses.

We've already explored some ways that our body is altered and transformed as it moves through different lands. If this is so, it is because the body is *itself* a kind of place—not a solid object but a terrain through which things pass, and in which they sometimes settle and sediment. The body is a portable place wandering through the larger valleys and plains of the earth, open to the same currents, the same waters and winds that cascade across those wider spaces. It is hardly a closed and determinate entity, but rather a sensitive threshold through which the world experiences itself, a traveling doorway through which sundry aspects of the earth are always flowing. Sometimes the world's textures move across this threshold unchanged. Sometimes they are transformed by the passage. And sometimes they reshape the doorway itself.

Despite the unending attempts to define and diagnose the body as a determinate object, the metamorphic character of our flesh makes itself evident in the most disparate contexts, whether when extricating oneself from the muddy suck of a swamp, or while caught up in the electric buzz and bustle of the city. Lest the experiences recounted in the previous chapter leave the impression that the capacity for metamorphosis is an entirely exotic affair, endemic only to peculiar persons dwelling in far-off mountains, allow me to provide a few very mundane examples, drawn from an ordinary life in North America, before taking up once again the matters that unfolded during my sojourn in the high valleys of Nepal.

The modern world of commerce and entertainment engages our corporeal susceptibility in multiple ways. I was in my late teens when I first became aware of the lingering influence that certain films at the local cinema had upon my organism. I am not referring here to the obvious trance we all fall into while immobilized before the big screen, but to the uncanny way that certain films would surreptitiously enter into my bloodstream, like a contagion. James Bond films were especially effective in this regard. When after the closing credits I filed out of the theater into the open air of evening, I'd naturally shift my thoughts to the practical matters of the moment. As I approached the corner of the block, however, trying to remember where I'd parked my car, my body would unexpectedly leap to the edge of the corner building and flatten itself against it, then slowly peer around the edge. Ascertaining that no one was approaching, I'd dash across the sidewalk to duck beside someone else's automobile along the curb. I would survey the street beyond by gazing through the side windows of that car, moving slowly to keep my body's contour from obtruding beyond the outline of the vehicle, and so from becoming visible to any eyes on the far side of the avenue. But from whom was I hiding? Whose presence was I hoping to glimpse as I slunk through the shadows, or later as I drove home, keeping the speedometer steady, slowing to stop at each traffic light with unusual precision, my peripheral senses keenly alert to the cars on either side of me? Who or what was I tracking? I had no idea. I hadn't really chosen to enact any of these instinctive behaviors, but simply became aware of them as they were happening, amused and slightly startled by the curious spell that my organism was under. After fifteen minutes or so, the weird veneer of secrecy tinting the whole of my perceptual field would finally dissipate, the possession of my bloodstream by a phantom agent of espionage now finally exorcised by the press of homework or taking out the garbage.

This curious capacity for being drawn, physiologically, into the terrain of certain stories—abducted into another landscape that would only belatedly release me back into the palpable present—would also show itself, now and then, in relation to certain books. Since I accomplished my most pleasureful reading in bed, late at night, the chemistry of sleep usually served to reorient my limbs and finally transmit me back, in time for breakfast, to the thick of the commonplace. But when I finished reading a fat, nineteenth-century novel not at night but on some morning in late summer, placing it carefully back on the bookshelf, the transition was not so smooth; present-day objects, like the disposal unit loudly masticating scraps beneath the drain in the kitchen sink, and a bus wheezing by on the street, all struck me as ludicrous anachronisms, unreal apparitions beamed in from another planet. The pattern of printed words in that novel had rearranged my neurons: time itself was out of joint, and remained so for several days.

But written texts held another, even loopier magic, a capability for metamorphosis that became evident only when I took up the craft of writing. It is a type of transmutation well known, I suspect, to many writers, although precious few probably notice how bizarre this phenomenon really is. It shows itself when, after laboring over some essay, I send a draft to a couple friends or colleagues in hopes that they might read it. Whenever one of them subsequently mentions (on the phone, perhaps) that she is now perusing the piece, then—at that moment—a curious change occurs in the text. When I glance at my copy of those pages, I find that every passage has been altered, some dramatically, some only slightly. It's the damnedest thing. But wait: looking closer, it appears that the precise words I've written are still the same, as is the order of those words, and the punctuation marks between them. Yet the whole thing reads differently. Problems I hadn't seen before are now glaringly apparent, while previously unnoticed elegances also reveal themselves, gleaming like half-hidden gems among the clumped grass of the text. For I'm now reading the text with eyes very different from my own. My colleague has not even called me back to tell of her reaction to my draft. In all likelihood, she hasn't yet finished reading it. It doesn't matter. I do not need to know if she likes or despises the thing, only that she is reading it. That knowledge, alone, suffices to activate a dramatically different encounter for me with what I've written. And if I learn that *another* friend is reading it, then the text will transform yet again.

Once I discovered this shapeshifting, polymorphous capability in my written work, I realized that as an author I'd be much better served if I could effect these transformations on my own, *before* sending the pages out to be viewed by anyone else. After finishing the first drafts of subsequent articles, I undertook to read them from the imagined perspective of various colleagues. But to no avail. I was simply unable to conjure, on my own, any clear change in the texture and feel of what I'd written. Only by sharing it with others did the metamorphosis occur . . .

Even at this very moment, then, your perusing of these printed words is already influencing them, changing their tones, torquing the way they feel to others, and to me.

Unless, that is, I'm simply more susceptible than other authors to such mutations in my texts. If so, this predilection is likely related to a porosity of feeling that has been with me since childhood, an impressionable bent that shows itself, often enough, in matters linguistic. During my sophomore year of college, for example, I first became aware of my unsettling propensity to inadvertently take on the accent of whomever I was speaking with. Whenever I was on the telephone, my housemates always knew the nationality of the person I was talking to. If my lips were pursed, they knew I was conversing with one of our French classmates, while if I was talking in gutturals from the back of my throat, it was obvious that I was talking to a Russian classmate. If my arms were gesticulating wildly, flying through the air around the phone as I spoke, it was a safe bet that I was talking to a student from Italy. When one of my housemates, laughing, pointed all this out to me, I was appalled. I'd had no idea! Surely, I suggested, everybody else does the same? He shook his head. And indeed, once I began paying attention, I found that the vocal inflections of most folks varied hardly at all when they spoke to different people. But in myself this mimetic proclivity was instinctive and unconscious; it was only with a great effort that I was able to interrupt such changes in my own inflection and accent. Unfortunately, such interruption also disrupted the easy flow of my conversation, making me trip over my tongue and get tangled in the words. I finally gave up trying. But it was disconcerting as hell to know that my voice so easily took on the attributes of those with whom I conversed. I wondered what circumstance had rendered me so darn spineless, so passively porous to the style of others.

Two subsequent discoveries finally reconciled me to this odd proclivity of mine. The first happened when, as an itinerant magician, I began traveling more widely, interacting more often with people from other lands. Even in North America I noticed that foreigners gravitated to me, finding me somehow much easier to communicate with than most of my comrades. This was not merely because my accent would shift in a way that made my voice easier to understand, but also because whenever I found myself in conversation with someone from elsewhere, all of the colloquialisms endemic to the American idiom simply dropped out of my vocabulary. This, too, was an entirely inadvertent and unconscious move. For I had no overt awareness of which terms and phrases were idiomatic, and which were more proper to English as a whole. I had all but failed my high school English classes, utterly unable to fathom the byzantine grammatical rules for the proper usage of this language. Of course, I had no problem *speaking* the language, and in fact speaking it quite well; I *inhabited* the grammar easily enough, I just couldn't make sense of it from the outside. So it was strange to discover that, without ever having paid attention to which turns of phrase were endemic to the local vernacular (having simply imbibed all these phrases at the same time as the rest of the English language, while growing up in suburban New York), my body nonetheless seemed to know, instinctively, how to make that distinction, and it spontaneously avoided such idiomatic phrases whenever it was conversing with someone from another land. I noticed this involuntary habit only by contrast, when I heard other friends talking to those same foreigners and was startled by the abundance of slang phrases and local colloquialisms that were unwittingly laced through their speaking, making it impossible for the exasperated visitor to follow much of what they were saying. But how could *I* so easily taste which turns of phrase were of local vintage?

I did not know. I knew only that I had a remarkable knack for making myself understood, and for understanding others, and this began to ease my dismay at the mimetic or copycat nature of my own discourse. My earlier difficulty learning foreign languages from textbooks had convinced me that I should forgo studying languages; I was now surprised by the ease with which I would pick up other tongues when traveling abroad and performing on the streets. Once I could hear the lingo all around me—could see the stance in which native speakers held themselves as they spoke, the way their faces folded when forming certain words—then I was able to feel my own way into the language as well.

Still, this physiologic sensitivity to others (this excessive somatic empathy) often left me confused when hanging out with clusters of people. Large groups were completely discombobulating: the divergent emotions of others picked up inwardly by my body were hard to navigate, and could precipitate an escalating sense of chaos within my organism. For this reason I studiously avoided parties—except those where there was an especially good dance groove, where the loudspeakers pulsed with polyrhythmic music that made me move. As long as I could lose myself in the rhythms, then the discordant sensations would dissipate within that wider reverberation. The other partygoers, with all their masked or bristling feelings, their sadnesses and their highs, would dissolve into the wallpaper as I was caught by this demonic oscillation rolling up through my legs and fountaining out through my arms and fingers. Dance was the improvisational ecstasy that dissolved the crossfire of other people's emotions.

I discovered the real value of my hypersensitive nervous system, however, only when I ventured beyond the confines of civilization, and began lingering in traditional indigenous communities. There, for the first time, I found myself surrounded by folks who assumed that every part of the land was alive. The feelings that played through these persons seemed less confined or bottled up within them, and more a part of the general circulation between cedar trees, humans, woodpeckers, mountains, and lizards, between cloudbursts and streams studded with trout. My porous nature seemed much less of a problem here. That permeability had always meant that I was too easily affected by other persons—by other nervous systems shaped so much like my own. Yet such porosity was just right, I now realized, for engaging nervous systems very *different* from my own. It was just right for entering into felt relation with other, non-human forms of sentience.

I simply had not noticed this gift back where I grew up, since other animals and plants were not acknowledged there as sentient beings capable of creative expression. But in traditional and tribal communities I found myself among people who practiced paying attention to the polyvalent speech of a landscape assumed to be quivering with intelligence, and who regularly turned to the sensitives among them to interpret the land's gestures. Here, such empaths were valued for their ability to dialogue with the leafing earth and its animal powers. The fact that a person was too impressionable to be at ease in the fuss and tumult of the human throng was a sure sign that she should make her home on the periphery of the village, working as an intermediary between the human settlement and the wider community of animate beings. Most of the medicine persons whom I met were precisely such individuals, whose sensitive nature empowered them to tend the boundary between the human collective and the local earth. By communicating (through their propitiations and their chants, through their dances and ecstatic trances) with plants, with other animals, and with the visible and invisible elements, the medicine persons' craft ensured that the boundary between the human and the more-than-human worlds stayed, itself, permeable—that that boundary never hardened into a barrier, but remained a porous membrane across which nourishment flowed steadily in both directions.

Sonam was just such a person, keenly empathic, given to living at a distance from other two-leggeds so he could more directly hear, and engage in, the layered conversations occurring between the milky ice-melt rivers and the rhododendron forests and the monsoon clouds creeping slowly up the valleys like glaciers in reverse. He struck me sometimes not as a person but as a walking piece of the mountainside, acutely sensitive to the grasses sprouting from his loamy surface, and to the animals, large and small, that moved among them, but also as implacable as the rock that underlay that shallow soil, and as inscrutable to me as the weather. Sonam had an especially close rapport with certain beings of the neighborhood, including a cool wind that poured down the slope behind the house shortly before each dusk; it issued from a single small pass in the high ridges, and Sonam would climb up there to immerse himself in that rushing intelligence. There were other familiars to whom he introduced me, and certainly many that he never shared. But Sonam was an especially close ally, and apprentice, of Raven. In this he was similar to his maternal grandfather—that deceptively immobile old man whom I'd watched endlessly flicking prayer beads in the corner of the first Sherpa household I'd slept in.

In the days after the thought-shattering transformation I reported at the close of the last chapter, when I saw Sonam turn into a raven and back again, I was able to piece together some elements of how, I suspected, that remarkable feat was accomplished. For only after that event did I become aware of how thoroughly Sonam gave himself to the patient observation of those black-feathered creatures, gazing silently at a group of ravens perched and arguing in a cluster of low trees, or at a single dark bird standing on a near rock. Before the transformation, I had sometimes seen him from behind when he was sitting facing the gorge, cross-legged and immobile on the ground as his head rolled slowly from side to side between his shoulders. I had assumed this was some kind of yogic practice for relaxing the neck and spine. Now I recognized that the slow swiveling of his head was actually accompanying the flight of a raven as it spiraled, riding a warm updraft along the cliffs to the upper slopes. Sonam turned his whole head rather than his eyes, his focus fixed on the bird's body as it banked and turned.

I discovered that he could sometimes call a soaring bird out of the sky, uttering a croaked cry with such precision that the raven would break its trajectory with a tumble and swerve onto the roof of the house.

It was not just the loud squawks that Sonam had perfected, but a whole lingo of conversational croaks and gutturals that I happened to have heard him deploy only once, when he thought I was not around, in interaction with three birds perched in the branches of a dead tree. The ravens had alighted in the snag while I was resting on the dirt with my back against a boulder, and soon began gabbing among themselves, clicking and aowawing and yawking, generating an impressive array of fairly soft, garrulous sounds I'd never really noticed before, and after a while another raven seemed to join the conversation from somewhere on the ground behind me. I listened for a time, and then peered around the rock: it was Sonam himself standing there, talking with the birds. As soon as he saw me he stopped. Now he was just listening to them. I bowed silently and walked away, wondering if he had merely been copying them or whether—although I could not quite bring myself to believe it—he was actually exchanging specific meanings with those birds, and they with him.

What was clear was that he had learned to precisely shape his throat and tongue in order to utter sounds that seemed, at least to me, perfectly indistinguishable from those of a raven. I recalled that just before I witnessed his avian metamorphosis, I had heard a raven's loud caw coming from around the bend in the trail. It was only then, as I rounded that bend, that I caught sight of the raven, which cawed at me twice before, well, transforming into Sonam. And so I reasoned that Sonam had waited for a certain moment along the footpath (a trail whose twists and turns he knew intimately) when he would be out of my view for a long stretch. After positioning himself, he had used the loud croak of a raven to set up an expectation in my organism—an anticipation in my eyes—that as I rounded the bend I would encounter a raven.

Of course, there was much more involved. As a student of ravens, Sonam, I'm quite sure, had long practiced holding himself in the various postures of that bird, had practiced Raven's ways of walking, of moving its head, of spreading feathered limbs. Learning to dance another animal is central to the craft of shamanic traditions throughout the world. To move as another is simply the most visceral approach to feel one's way into the body of that creature, and so to taste the flavor of its experience, entering into the felt intelligence of the other. I have witnessed a young medicine person in the American Southwest summon the spirit of a deer by dancing that animal, have watched a Kwakiutl magician shuffle and dip his way into the power of a black bear, have seen a native healer dream her body into the riverdance of a spawning salmon, and a Mayan shaman contort himself into the rapid, vibratory flight of a dragonfly. In every case, a subtle change came upon the dancer as she gave herself over to the animal and so let herself be possessed, raising goose bumps along my skin as I watched. The carefully articulated movements, and the stylized but eerily precise renderings of the other's behavior, were clearly the fruit of long, patient observation of the animal other, steadily inviting its alien gestures into one's muscles. The dancer feels her way into the subjective experience of the other by mimicking its patterned movements, and so invoking it, coaxing it close, drawing it into her flesh with the subtlest motion of a shoulder, or a hip, or a blinking eye.

A key element in such kinetic invocations of another animal is the magician's ability to dream himself into the wild physicality of that Other, allowing his senses to heighten and intensify as he becomes possessed by the carnal intelligence of the creature. The shaman himself must be convinced of his transformation if we who watch are to have a chance of experiencing the animal's arrival. There is no room for fakery or mere illusion here; if the magician does not feel himself undergo a full metamorphosis into the other, then we who watch will never be convinced of the change. Yet this is not to say that there are not specific techniques employed to loosen our senses, particular perceptual methods used to enhance the invocation of the animal. Merely calling to the creature in one's imagination will never suffice; one must summon it bodily, entering mimetically into the shape and rhythm of the other being if the animal spirit is to feel the call. One must unbind the human arrangement of one's senses, and those of any humans watching, if the animal is to feel safe enough to arrive in our midst.

And so I was sure, now, that Sonam had timed his transformation along the trail so that it took place close to dusk, a moment when our eyes are less certain and more apt to confuse things. Further, he knew precisely at what distance he'd have to situate himself in order to appear, from that bend in the path, like a much *closer* figure about fifteen inches in height. He may well have chosen in advance the very boulder on which he'd stand, a rock whose plain surface, seen when rounding the bend, would allow it to appear as a much nearer and much smaller rock than it was, with not a man but a raven perched upon it. Sonam had waited till he was well out of my sight around that turn in the trail, had climbed onto that boulder and faced the path, bending his knees sharply so only the lower legs were evident. He had leaned his torso steeply forward while extending his arms straight back alongside that torso, his wrists and straight-fingered hands jutting past his rump like folded wingtips, entering the feathered mind of the raven by dancing that long-beaked form. He squawked very loudly as he heard me approach the bend, and then again LOUD, forcing the expectation in my organism that I was about to meet a raven at close quarters.

As I rounded the bend he simply kept up his dance, hopping on both legs together as he turned, swiveling his head in jerking movements, blinking his eyes like shutters and opening his beak to squawk one last time before hopping to the ground. During that descent to the ground Sonam lifted his arms and came out of the trance, or rather *I* started to slip out of the trance, for the way in which the raven dropped seemed incongruous. As my brain worked to make sense of what it saw, it first concluded that the raven was somehow much LARGER than a normal bird. This sudden growth in perceived size made it seem that the bird was swooping rapidly toward me (for such is often the case when a thing appears to grow rapidly larger), and so my hands flew up in front of my face. But then as it landed on the ground the sound was all wrong, somehow, and what had been a much-too-large bird resolved into a much-too-small person, until I realized that it was Sonam standing there, though much farther down the trail than the bird had been.

It had taken some time for my senses to recalibrate themselves. The shock of the encounter, my first wholly conscious witnessing of a full-bodied metamorphosis, was extreme. Certainly the transformation was made possible by Sonam's rapport with the ravens. But it was also enabled by a strange contortion of spatial depth—by a temporary reversal of near and far along a precipitous trail in the mountains. The metamorphosis was activated by a momentary slackening of the grip that my eyes and ears commonly have upon the space around me. A momentary derangement of my senses, provoked by the precisely timed utterances and antic contortions of Sonam's body.

And so not merely a confounding of *my* sensory organization, but an alteration of Sonam's as well. In order to take on the attributes of Raven, in order to feel the hollowing out of his bones and the feathers sprouting from his flesh, Sonam had necessarily to alter his own organization. Only thus would he have been able to discover the flight muscles within his breast, and the precise posture for his head and neck. I believe it was his thorough immersion in the experience that so completely compelled my own participation in the metamorphosis.

Sonam and I gently but carefully avoided speaking directly of the event. It had happened; this I knew, and he knew that I knew. The transformation on the high trail had clearly been a demonstration of sorts, but it was also a lure, a suggestion of skills to be attained, a visual conundrum that served to clear my mind of distractions and train my attention on the simple tasks that Sonam now set for me.

These tasks mostly took the form of perceptual exercises that Sonam instructed me to practice in between various daily chores. The most consistent of these involved sitting or squatting on the ground while steadily focusing my eyes upon an arbitrary spot on the near surface of one of the huge rocks in front of his house, and doing so for long stretches without wavering, and preferably without blinking. This was easy enough for a few minutes, but then increasingly difficult; my eyes would begin itching, and fill with tears, and so I'd have to blink if I wanted to see anything at all. After a single blink I'd keep them focused there, at that spot composed of flecks of gray and black and tan and silver, gazing and gazing, until the surrounding surface of the stone appeared to melt and started to writhe, while the still point where I was focused maintained its circular quietude, a calm pond in the midst of what now seemed a seething nest of serpents. A series of blinks, or a minute shift of focus, would suffice to return the rock to its solid composure. I kept gazing. Sometimes the rest of the surface would stay stable, but the very patch where I was focused would begin to give way, dissolving backward, it seemed, into the interior of the boulder, and my focus would lose itself in that molten thickness.

Sonam seemed aware of the different instabilities that the exercise would bring on; after a while he began to enhance these odd effects by instructing me to place my focus not on the surface of the boulder but *inside* the rock, at a point about a foot behind the spot I'd been staring at. This was far more challenging than the earlier task. (Such exertions, however, were not entirely unknown to me. The challenges posed by Sonam were at first analogous to visual experiments I'd undertaken some years earlier, in the course of developing various sleight-of-hand effects. Yet his exercises quickly pushed my sensing organism far beyond the limits of my earlier practice.) Several afternoons later, Sonam instructed me to fix my gaze at a point in the air midway between myself and the boulder—a spot he indicated first with his forefinger, so I could train my eyes upon his fingernail before he withdrew his hand, leaving me to maintain that focus, as best I could, for an hour or more. This was plainly impossible. Like the previous exercise, focusing inside the rock, it amounted to letting two flat images of the rock's mottled surface float off there in the near distance, like immaterial phantasms, while steadily resisting the impulse to resolve them into a single solid surface. It was a nerve-racking challenge, endlessly frustrating and headache-inducing. Nonetheless, I diligently brought my awareness back, and back again, to a point in the unseen air between me and that boulder (often imagining that a tiny insect was hovering there), and within three or four days I had developed a knack for keeping my focus suspended there, at least for a few minutes at a time.

The next directive was especially confounding. Sonam asked me to gather both of my listening ears into that small point in the air where my eyes were focused. What?!? I could not ascertain what he was asking until he acted it out with his hands: just as our two eyes regularly come together in a single focus, so also our two ears often converged upon a single point of sound out in the world—a yak bell ringing, a child crying. Sonam was simply asking me to concentrate my listening upon the very location where my two eyes were already focused. (I have since practiced this in crowded restaurants in the West, trying to pick out and listen in on a particular conversation within the general clamor of voices. To my surprise, I discovered that this was indeed much easier if I focused my eyes upon the specific couple whose precise words I wanted to hear. I did not need to see their lips, only to fix my gaze, and hence my awareness, upon their table.) This was no mean feat, given that there was no sound whatsoever coming from that point in the air between my face and the boulder—but once again I kept at it, trying to hone my concentration, letting ambient sounds fade into the background as I tuned the focus of my listening ears, always hoping that I might actually hear something at that spot if I listened intently enough. But I never did hear anything.

After fifteen days of such exercises, my muscles stiff from the long periods of immobility, Sonam took me hiking in the late afternoon up the slopes behind his home, switchbacking this way and that as the house slowly shrunk below us. After a while the grade became too steep for our legs alone; I had to use my hands as we climbed up between the jagged facets, finally reaching a small piece of level terrain—a narrow shelf beneath high cliffs. Through a break in those cliffs I could see the uppermost part of a single, resplendent peak to the northeast, its snowfields still agleam in the rays now long gone from the valley below us. We sat and waited in the crystal silence. Soon a breeze, quiet and cold, began pouring through that cleft in the rocks, and my friend began speaking a prayer to the moving air. The breeze became stronger. Sonam had me stand and face into it, as with his cupped hands he poured the gusts over my face, and then cupped them over my chest and stomach and legs. He had me turn, then, and directed the wind against the back of my head, onto my shoulders and my arms and the back of my legs, praying aloud the whole time. I understood nothing of what he said, and had no need to.

We made our way down slowly, with Sonam showing me where to place my feet. After a time, the wind fell silent and stopped. We reached the house as the dark itself settled around us.

On the next day, Sonam changed the exercises. I was no longer to gaze into boulders, but to train my vision—gently but unwaveringly—upon ravens. What if there were none around? No matter: I was to wait for them, to stay attentive to them, to be ready. While a few individuals seemed to hang out often enough near Sonam and Jangmu's home, there was a particular place that Sonam showed me, less than a mile along the slope to the southeast, and just a short way above the main footpath, where the valley's ravens liked to stop and linger. It was at the edge of a small patch of forest with assorted large rocks strewn about nearby, affording me several good sitting places, as the complexly bent and spreading branches of the low trees afforded perching for the ravens.

As before, it was a matter of the precise confluence of my eyes. Sonam asked me to try to focus upon a point just beneath the head of a raven, between the shoulders of the bird (he showed me on his own body) or, if the bird was facing me, at the top of its breast. But the ravens there, I found, rarely perched very long in a single spot before flapping to another. Still, the instruction was to keep bringing my attention back to that spot on each bird's body, until I found myself gazing at an individual less nervous or ready to move, a raven that was more relaxed in the moment of our meeting. When such an opportunity arose, I was to sink my focus as best I could into the body of the raven, to a point midway between the top of its breast and its shoulders (near the base of the bird's wishbone) and hold it there.

Sometimes there were no ravens present; more commonly, there were several haunting that forest edge, but at a distance of twelve or even twenty feet from where I sat. So this sinking of my focus into the body of a bird was a mostly imaginative act; the ravens were at first too far away for me to register or physically feel such a slight shift in the focus of my eyes. But Sonam was after something specific: he wanted me to feel the experience of meeting up *with myself* inside the bird. To feel the two sides of myself joining up with each other over there, in the torso of the raven. Although he spoke in simpler terms (constrained in part by my modest vocabulary), it seemed that Sonam was inviting me to notice the left and the right sides of my sensorium meeting up with one another over there, outside of myself, at that location in space where the separate gaze of my left eye and that of my right eye converged into a single focus.

The more I practiced this unorthodox meditation, the more I was able to sense what Sonam was after. If, while I was gazing one of the ravens, another swooped down and alighted somewhere *between* me and the bird I was watching, that interloper would hardly be noticed by me, or would be felt only as a vague ghost hovering between the solidity of my person and the solidity of the raven upon whom I was focused. It was as though I were no longer entirely located over here, where my body was sitting, for some piece of me had also gathered itself over there, beneath the purple sheen in the night-black feathers of the bird.

There was one male, however, who displayed a greater audacity than the others, swerving over to ponder me from close by (whether looking down from a near branch or peering sideways from the ground or—after a few days—from a perch on the same rock as I, although always just beyond the reach of my extended limbs). I at first thought there were several such individuals winging close to feed their curiosity, but soon realized that it was always the same fellow. If he was in the trees when I arrived in late morning, he'd soon swoop out of the leaves to look me over at close range. If he was out and about, then sometime later he'd likely drop in from elsewhere and bank onto a neighboring rock, squawking a few times, his clawed feet turning first one way and then the other to scrutinize me from each eye in turn. If I tried to utter a ravenlike croak he'd answer me back straightaway, as if correcting my diction, yet he was still more garrulous when I spoke to him in Sherpa or even English. I got to thinking about how, in many species, there are certain individuals who stand out among the others of their kind for their curiosity and cleverness.

Many months earlier, in a village near a wild forest preserve on the south coast of Java, I'd been warned by the local fishermen that there was an unusually bold individual among the bands of monkeys that roam the forest canopy, a particular monkey much more daring and skillful than the others, especially at stealing things from humans. Because I wear glasses, I was urged by the fishermen not to enter that forest, for that sly monkey was known to silently accompany people in the branches far overhead, waiting for an opportune moment to swing low and snatch the glasses from their face. The villagers had had to organize several search parties for missing visitors who turned out simply to have been wandering half blind for several days, unable to find their way out of the woods.

Similarly, when I lived in the northern Rockies there was an old bull elk who was legendary among the local hunters. Larger than the other males thereabouts, he had once had the biggest rack of any bull in those mountains, although in recent years (folks said) his antlers were smaller. He was glimpsed often, yet no one had ever succeeded in planting a bullet anywhere on his person. His ability to elude hunters was uncanny, enabling him to melt away and vanish even as the hunter registered the glimpse. In earlier years the locals had taken the large bull's readiness to show himself as a challenge, with each hunter eager to finally shoot him and be able to boast about the fact. But after so many years, the old one's continued defiance of hunters had made him not only a legend but a revered spirit among the hunters and everyone else in the region. Hiking one October evening with a friend who'd grown up in that area—and hearing now and then the most beautiful of all earthborn sounds, which is the autumn bugling of elk—there abruptly sounded from far off the most heart-wrenchingly lovely of any call I'd ever heard, a bugling that was the most full-throated and deep and at the same time the most ethereal, ascending slowly upward through a sequence of clear overtones before ending in a series of gutteral grunts. I looked wide-eyed at my friend. It was the unmistakable call, he said, of that great elder, the phantom.

In these cases, and I could mention many others, the uniqueness of the individuals seems to reside not just in their intelligence but in their skill at interacting with other species. Since we notice their uncommon savvy in their dealings with *us,* we might assume that these animals display such chutzpah only toward humans. But this seems unlikely. That old elk doubtless relies on his remarkable wiles in relation to other predators as well, and (I can't help but suspect) in his relation to every aspect of those wooded slopes, to unexpected changes in the seasonal cycle, or the sudden arrival of roads, and clear-cuts, in a favorite part of the mountains.

The observation by indigenous peoples that there exist particular individuals—among other animals as among our own two-legged kind—who are in a strangely different league from their peers has led some native traditions to posit that there exists an entirely different species to which such individuals belong, a class of entities who are able to cross *between* diverse species, taking on the ways of various animals as needed—able to trade wings for antlers, or to forsake paws for scaly fins or even fingered hands. This is the class of those who are recognized, when they're in human form, as shamans—as magicians or sorcerers. But most contemporary persons, lacking regular contact with the wild in its multiform weirdness, have forgotten that such shamans are to be found in *every* species, that in truth they are a kind of cross- or trans-species creature, and hence a species unto themselves.

There was an odd thing about this one corvid whose curiosity led it to settle closer to me than the others dared. Its proximity, of course, made it the best possible subject for the practice Sonam had prescribed, fixing my gaze as best I could upon the feathers just below the bird's head and then, once it was stabilized there, sinking my focus into the center of the bird's flesh. Well, that raven had no problem with my gazing back at it, concentrating my eyes upon its ruffed feathers. But whenever I tried to accomplish the last step in the exercise, letting my attention penetrate behind the bird's feathered surface, it would squawk and flap away to some farther vantage, as though offended by my intrusion. When, shortly after, the same bird would swoop back to gawk at me from a near perch, it took time for the two of us to reestablish some sort of basic trust as we pondered one another. Still, whenever I tried to move my focus into its body, the raven would get distressed and fly off.

It seemed hardly possible that the bird could notice the shift in my attention, for there appeared no movement on my part, only a tiny change in the angle of my eyes, while the larger changes in my focus (accompanying the bird as it moved) seemed not to faze it at all. Nonetheless, it happened again and again whenever my gaze dropped deeper; clearly I was violating the etiquette between us, and so I was never able to complete the exercise with this particular corvid. I finally gave up and started paying more attention to the other individuals.

Sonam came by one afternoon to check on whether I was making progress with the ravens. I had told him, the day before, that I was unsure whether I was really able to move my focus into a bird's body, but that I did have the *feeling* that I was doing so. So he came 'round to watch and feel for himself whether I was getting anywhere with all this. He must've been satisfied, for he now added a further step to the exercise. I was to bring another sense, specifically my *tactile* sense, into the juncture where my eyes were focused.

It took me a while to fathom what Sonam was instructing. He first indicated the sense of touch by moving his fingertips across a rock's surface. When I concluded that he was talking about the fingers, he said that was not what he meant. He began pinching the skin of my arms and legs and face until I finally got that he was talking about the sense of touch in its entirety. He wanted me to combine feelings of touch with the visual sense, or rather to bring my tactile sense to bear, over there, where my two eyes converged into a single focus. This became clear only when Sonam offered the example of looking into the fire at his house: if he just gazes at it with his eyes, then the flames will be interesting to look at, but when he adds his tactile sensibility to the focus of his eyes, then while watching the flames he will soon—as he mimed with his body—come to feel more and more hot inside his chest. He then led me through the woods to where a small brook gurgled its way down the slope. Here, he said, if he looks at the brook with his eyes alone, then he can see all sorts of things, including the surface patterns and the pebbles on the bottom, yet he feels nothing different inside his body or along the skin of his arms. However, when he brings the sense of touch into the focus of his eyes, then (as he again acted out for me) he begins to feel the fluidity of the water moving within him; it cools his insides and eases any tightness within his muscles.

From these demonstrations I gleaned that for Sonam, the tactile sense comprised not just the surface sensations of his skin but also all the visceral sensations within his flesh. I tried to feel something of what he was talking about while gazing into the cascading brook, without success. No matter. It takes much practice, he said, and will be easier if I practice with the ravens, with whom I was already getting somewhere.

So: while watching the corvids I was to marshal my tactile sensitivity and join it, somehow, to the focus of my eyes. A tall order. But I poured myself into the effort . . . with negligible results. It was rare enough that I could settle my gaze on a bird whose stance was quiet and stable. But then whenever I tried to bring my tactile awareness into that focus, the *here-ness* of my various body sensations would break my concentrated attention *over there,* unbuckling my eyes from their grip upon the bird, and so it would take me a while to find that focus again.

You might think that pondering these dark birds day after day would be boring. But this was hardly the case; ravens are outlandishly interesting creatures, graceful and awkward at the same time, cantankerous, keenly responsive to one another and to events in their vicinity. They are stunningly skillful fliers, as compelling on the wing as any raptor, yet apt to break into crazy looping stunts, sometimes in tandem with one another. The ravens in that valley had a wide range of cries, some grating and some bell-like and smooth. By lending them my attention I discovered as charismatic a creature as can be found carousing anywhere on earth.

After several days of exasperated effort spent on the baffling task set for me, the fruition arrived unexpectedly, when I'd given up for the afternoon and was making my way back toward the hut. A couple hundred yards along the trail I came upon a raven crossing the dirt to peck at the corpse of a small rodent. As the bird leaned forward, I felt something inside me tip forward as well, and lost my balance for a moment. I regained my equilibrium as the bird kept pecking at the carcass, but now couldn't help noticing a sensation in my neck every time the raven reached its beak toward the ground. After a few tries, the bird succeeded in loosening a large morsel from the remains, and swooped up onto a shelf of rock with the gore in its beak; as it did so I felt a sudden weightlessness in my chest, which abated as the raven settled onto the ledge. Had I really felt that? Yes!!! I knew immediately that this was what Sonam had been nudging me toward. The sensations were subtle, but unmistakable. As if the bird outside me had somehow awakened an analogue of itself inside my own muscles. Or, rather, as if the raven were not only pulling apart that bit of blood and meat out there on the rock ledge, but was also doing so in here, within my own organism.

In truth, I was not sure whether I had induced, by my exertions, a new set of impressions within me, or whether I'd simply become awake to a range of visceral sensations that had already been present below the threshold of my consciousness. For I soon noticed such inward sensations accompanying most of my outer, visual experiences—whether I was watching trees flexing under the press of a strong wind or a boy bent low beneath a load of wood as he climbed barefoot up the trail, or water buffalo moving through the terraced fields, goaded along by a little girl with a branch twice as long as her own body. In every case, their outward movements coincided with faint sensations within me—dimly felt tugs, torsions, and twists within my sensorium echoing the visible tensions and loosenings in the bodies around me. A field of wildflowers, too, could be inwardly felt if they were moving in the breeze, and now even the small white cloud huddled atop a single peak across the valley was, whenever I glanced at it, a kind of coolness on the back of my neck. The next morning, that same cloud was blowing off to the eastern side of that peak, except that it never dissipated, billowing there all afternoon like a flag—the chill no longer centered on the back of my neck but rippling across my left shoulder.

Was this, then, the truth of perception—the body subtly blending itself with every phenomenon that it perceives? During those days, it began to seem as though my body was not, properly speaking, mine, but rather a piece of the sensuous world—and *seeing* was a steady trading of myself here with the things seen there, so that this sensitive flesh became a kind of distributed thing, and the visible terrain a field of feeling. And yet, as I noted—scribbling—in my journal, there was still distance and depth. The commingling of myself with things did not dissolve the distance between us, and so the sentience at large was hardly a homogeneous unity or bland "oneness," but was articulated in various nodes and knots and flows that shifted as I moved within the broad landscape: that round rock overhanging the cliff's edge feels like the right knee of the valley, as that jostling bunch of trees across the river far below seems an agitation within the groin of the world, and the ribbon of water way down there is now, yes, a thread of icy clarity winding up my spine.

Perception alters, and with it the earth. The magician's body is a kind of cauldron brewing potions that alter their powers according to the precise blend of senses involved; he offers these in turn to his apprentice, whose creaturely body slowly awakens, loosening itself from societal, fear-induced constraints.

It was through Sonam's tutoring that I came to recognize the astonishing malleability of my animal senses. During those weeks in his valley I discovered these simple truths about perception: that neither the eyes, nor the ears, nor the skin, tongue, or nostrils ever really operate on their own; that each sense is steadily informed by other senses; that as we explore the terrain around us, our separate senses flow together in ever-shifting ways.

Neuroscientists give the name "synaesthesia" to the blending or coalescence of different senses. Synaesthetic experience is often studied as a *confusion* of the nervous system, and has commonly been viewed as a kind of pathology. *Synaesthetes*—that is, persons beset by this supposed affliction—are liable to *hear* the sound of certain colors, or *taste* the flavor of certain sounds. Our civilization prizes analytic precision, breaking things down into their component parts; it becomes frightened when things refuse to stay separate, when a person's senses flow together, so that her entire body begins to function as an interrelated organ of perception, as a single, complexly gifted sentience.

As a conjuror I had long been fascinated by the mutable character of perception. Back in college I had even written about the way the practice of sleight-of-hand purposely confounds the conventional segregation of the senses, making use of a synaesthetic propensity much more prevalent than science assumes. In the mountains of Nepal, however, I came to recognize that synaesthetic experience is not just commonplace; it is the very structure of perception. In our encounters with the world, our senses steadily intercommunicate and meld. Only by conjoining different sensory modalities can our organism garner insights into the specific otherness that confronts us at any moment.

Suppose that you are out strolling near your home, and a neighbor's dog comes bounding up, eager to be scratched under his collar and to chase a stick that he hopes you will toss. In order to do so, you must wrestle the stick from his slobbering mouth, a matter that can be accomplished only with a great amount of tugging to one side and yanking to the other, with the mutt yanking back, growling in the back of his throat while keeping his teeth locked upon the wood. Soon he will probably let you have the stick, but for now the gusto of grappling with you is irresistible, and the hound is putting up a good fight.

At such a moment, you do not have distinct experiences of a visible dog, an auditory dog, and a tactile, furry, saliva-dripping dog. On the contrary, the dog is precisely the place where those divergent senses link up and dissolve into one another, merging as well with the dusky odor ghosting around him like a cloud. Perception *is* this very commingling of different senses in the beings we perceive.

To simplify matters for a moment, let's consider only the sense of sight. Even here, within a single sense, we have not one but *two* eyes—*two* organs, each with its own angle of vision. Yet we don't commonly experience our left eye's perspective on, say, a clump of moss as distinct from our right eye's perspective on that moss. No, rather, that clump draws the two gazes together; the separate perspectives of our two eyes converge and merge in that emerald softness, and by virtue of their collaboration we glean a sense of the thickness and depth of the moss. Hence, the simplest act of seeing is already a kind of synaesthesia, folding two gazes (from two discrete eyes) into a single vision.

Now, just as a phenomenon seen by only one eye lacks apparent depth, so an object that impinges upon only a single sense—provoking neither the participation nor even the imagination of the other senses—remains a vague and insubstantial presence for us. It will gain in substance and reality to the degree that it tempts the involvement of one or several of our other senses. The faint, high-pitched hum of a mosquito near our ears may be sufficient to set our arms swatting at the air, yet this is because the sound already provokes a tactile association that distresses us. If we manage to catch a glimpse of that whirring sprite as it weaves through the space around us, then the mosquito will acquire a more vivid and precise reality, and if it succeeds in drawing a bit of blood from our ankle or buttock, then the tangible sting will combine with that sound and that blurred motion, yielding a still stronger sense of this tiny being as a keen and willful adversary—as another shape of intentful life.

Strolling through Manhattan in the winter, listening to Marvin Gaye on my MP3 player, I stop to watch a graceful figure skater cutting sinuous patterns at the outdoor rink. With my headphones on, the skater seems a pretty piece of the scenery, nice to notice but basically just part of the view. When I slip off the headphones, however, the audible scrape of her blades fuses perfectly with the visible spray of fine powder as she swerves and whirls, and so I'm tugged into visceral relation with the skater. The conjunction of my listening ears with my gazing eyes induces a subtle collaboration of my body with hers. So when the tip of one blade catches unexpectedly on the ice and she teeters, her limbs akimbo, before her shoulder slams down against that rock-hard surface, I feel the abrupt disequilibrium and then cringe as the ice seems to collide with my own bony shoulder, sending waves of excruciating sensation radiating across my torso before they subside into the sight of the skater climbing to her feet.

The empathic propensity of our body is in large part a consequence of the differentiation and divergence of our several senses. For it's only by turning our bodily attention toward *another* that we experience the convergence and reassembly of our separate senses into a dynamic unity. Only by entering into relation with others do we effect our own integration and coherence. Such others might be people, or they might be wetlands, or works of art, or snakes slithering through the stubbled grass. Each thing, attentively pondered, gathers our senses together in a unique way. This juncture, this conjoining of divergent senses *over there,* in the other, leads us to experience that other as a center of experience in its own right, and hence as another subject, another source of powers.

Incomplete on its own, the body is precisely our capacity for metamorphosis. Each being that we perceive enacts a subtle integration within us, even as it alters our prior organization. The sensing body is like an open circuit that completes itself only in things, in others, in the surrounding earth.

The bird walks forward, swaying from side to side with each step, then stops, cocking its thickly feathered head to watch another raven flapping by. It tilts its head the other way, caught by the rasping sound as Jangmu begins grinding chili peppers between two stones, then turns its whole body to peer, it seems, at me. I am gazing intently, not at those dark eyes but at that point just beneath the ruffled neck feathers. The raven knows I am watching, and perhaps can even feel the grip of my gaze within its chest. Soon, however, its neck feathers smooth back down and I sense it relax. It can hardly be very agitated, for it has not ventured beyond this small patch of ground since alighting here over half an hour ago.

Sonam had pointed out this bird to me when it swerved above the house. "It will happen today," he said.

We had just carried a long piece of deadwood from the main path down to the stone house, to be carved into a thin bench. "What will happen?" I asked.

"That is the one you must watch this morning," he said. "Wait until he finds a spot on the ground. Don't sit too close." He gave me one or two other instructions while the raven made a wide circle above the boulders. Then the bird, with its wing feathers splayed, banked into a landing on the bare ground close to the edge of the gorge.

I wandered over quietly, slipping past the clustered boulders and crouching down about six yards from the bird, facing it, with the edge of the canyon off to my right. My right leg was folded beneath me, my right buttock pressed into its heel, while my left knee was upright before me with my left forearm resting crosswise upon it. I keenly remember this asymmetrical posture, for it has become, since that day, my preferred position for grounding myself and drawing clarity to meet the contortions of the world.

Sonam had warned me not to look directly at the bird's eyes, but to sink my visual focus into that place within its breast, and to hold it there. I had been doing this for a long time without altering my position, my eyes fixed steadily upon the raven as it walked and as it rested, peering this way and that. Now it leaned to peck at the dirt—seeds, I thought at first, although when it straightened up I saw that a large beetle was in the beak, which quickly opened and snapped shut around it.

"You can taste what he is tasting?"

Sonam's voice was soft beside me; he'd come up without my noticing. He spoke again: "You can taste that bug as he swallows it?"

"No, I can't," I said.

"Yes, you can," he replied. "Try."

There was a gritty bitterness to the saliva in the back of my mouth as I shifted my attention there. I tried to sense what the raven might be tasting, letting my eyes close a bit as I did so.

"No! Keep your eyes open!" whispered Sonam. "Let them meet inside the bird. Hold them there."

"Okay, yes."

Sonam was now on both his knees beside me, his torso leaning forward, staring—like me—at the night-black creature.

"Now bring your touching to where your eyes meet."

"Like before," I whispered.

"Yes. Move all your feelings into the bird."

Once again, I tried to marshal my tactile and visceral sensations into that place, over there, where my eyes were focused. This was tougher now, for I was aware of Sonam next to me and felt very self-conscious. But then I noticed Sonam's hand on my left shoulder, his fingers pressing into my flesh.

"Move into the bird," he said, and I heard his voice as if at a distance. "Keep your eyes open. Eyes open. Watch."

The bird is now hopping, not walking, toward the edge of the gorge, and I feel each hop as a slight jolt. Its shoulders expand as wings spread and lift, and then with a lunge we are aloft . . . The rim passes under us and the ground falls away into a terrifying emptiness, as the whole canyon opens beneath us. The cliffs just below are glinting silver and pink in the sunlight; they descend in ledges and shelves that expand as we swerve toward them and then tilt away from us into the sky. The horizon angles up sharply and a series of vertical crevasses open and close like pages of a book as we glide past, and then the immensity of the canyon yawns around us with the rush of the rapids far below. A vertigo rises from my belly into my throat and I'm falling, I'm Falling, Gonna Die For Sure, but suddenly hear this voice—"Eyes Open!"—and feel Sonam's fingers pressing harder into my left shoulder, and I stretch open my eyes to see more. Now we're following the blue ribbon of water as it gets bigger and wider and *louder,* its many voices swelling as a freshness fills the whooshing air. I'm gulping for breath as I watch water charging over rocks and spilling around boulders, splitting and rejoining itself over and again, but now we're swooping faster than the river's flow; the green of a few leaves and then whole trees heaving and sighing just beneath us as we turn (can this be that thicket of rhododendron trees way downriver?) and it feels like we're gonna land here yes oh please yes please please yes but we tilt away and now two other black birds are flapping up outa those trees, calling back and forth. Then cliffs are slanting past and the river's falling away, and then the cliffs close by again, then the river, then the cliffs, then that abyss, and I finally realize we're spiraling up the side of the canyon, riding one of the warm updrafts like I've watched ravens do so many times. Soon we're above the main gorge, rowing up through air's thickness into a small valley without trees, then tilting out of the valley, flapping hard to rise above a ridge, the air humming around us as we stroke toward the shadowed face of a pyramid that grows rapidly larger and larger, ice fields in its furrows and dark deltas of tumbled rock as this black wall fills our sight, its upmost angles blurred by mist, until I recognize by some detail the same banner of cloud that I look at every day from far across the canyon, streaming off the top of this peak. We're banking in a wide arc toward the right, and then the vision that opens beneath us shudders through the breast and backbone: glacier-hung mountains upon mountains spreading off into the white distance. A gleaming, glistening world without people . . .

And I'm balancing, floating, utterly at ease in the blue air. As though we're not moving but held, gentle and fast, in the cupped hands of the sky. Stillness. Through a tangle of terrors I catch a first sense of the sheer joy that is flight. Falling, yet perfectly safe. Floating. Floating at the heart of the feathered thickness that is space. Aloft at the center of the world mandala, turning it beneath us, the whole planet rolling this way or that at the whim of our muscles.

The cliffs at the far side of the gorge now rising to meet us as we dive. Among the rocks scattered near the chasm's edge there's a rectangular boulder we're falling toward. A wisp of smoke is rising from that boulder, evident only as we bank above it. The thought comes that that rock is maybe a house. And there, off past the other rocks toward the edge of the precipice is an odd creature—no, two creatures, two clothed people crouched together on the ground. Their faces are upturned, staring steadily at us even as we glide downward, their heads turning together as they track us perfectly with their gaze. The eyes of one are especially compelling, achingly so, staring straight toward, straight up into . . .

me.

THE REAL IN ITS WONDER

(Language III)

We know what the animals do, what are the needs of the beaver, the bear, the salmon, and other creatures, because long ago men married them and acquired this knowledge from their animal wives. Today the priests say we lie, but we know better. The white man has been only a short time in this country and knows very little about the animals; we have lived here thousands of years and were taught long ago by the animals themselves. The white man writes everything down in a book so that it will not be forgotten; but our ancestors married animals, learned all their ways, and passed on this knowledge from one generation to another.

—A CARRIER INDIAN,
FROM BRITISH COLUMBIA

For how many hours have I been stumbling through this folded terrain? Pink spires leer down at my tiny form as I walk along the edges of gullies, slipping into and out of fractaled ravines. Stratified bands of pink and gray snake along the escarpments, maintaining their respective thicknesses as they round the folds and the fissures. No real shadows in this striated cosmos—no trees to speak of, only a few shrubs dotting the landscape in those spots where scant rainwater swerves into rivulets when a rare thunderstorm bursts through the monotonous heat, cracking the skin of the earth. The French trappers called this realm *les mauvaises terres à traverser,* "the bad lands to cross through," while the Oglala Sioux called it simply *mako sica,* "bad lands." Yet for eleven thousand years Indians came to spend time in this place, most often to hunt the various creatures that took refuge here, and now and then to lure their enemies into these furrowed labyrinths and so to confound their wits. Arrowheads and other hunting tools are steadily eroding out of these old cliffs, along with charcoal from countless campfires. Deeper strata give up fossils of ancient mammals, fossilized turtle shells, and sea shells. The Indians, coming upon these traces from the Eocene and Oligocene epochs, rightly assumed that this land was once covered by ocean. The old oral traditions of the Sioux and the Arikara peoples show their understanding that those ancient waters, and the land around them, were once populated by large creatures that no longer existed, beings whose bones were nonetheless left in the layered rock of those parched gullies.

I've seen no trace of humankind, however, just a jackrabbit darting between the bushes, and what I think was a kestrel hovering against the cliffs.

I left the car by the side of the highway this morning, drawn out of the vehicle by fascination with what my eyes were seeing. I slung a strip of cloth dangling a bottle of water over my shoulder and wandered out into this impossible landscape. The longer I wandered (my legs following the directives of the riven ground), the wider the scape seemed to grow. By noon I was thoroughly, blissfully lost in the sculpted immensity of this place—a small two-legged insect making its way between the toes of giants. Each corner that I peered into seemed a world in itself, and if I ventured in to explore further, that nook would expand around me and I'd find myself enveloped within its intimate expanse, held within the hugeness of a realm composed of worlds within worlds within worlds.

How hopelessly insignificant the human is—a speck lost and blundering through the limitless depths of the macrocosm, drinking from the voluminous air, dazzled by reflected rays and drawn upright by a hundred flesh-toned cliffs and slopes.

Such encounters with the outrageous scale of the larger Body we inhabit bring a shuddering humility, yet they can also release an unexpected intuition of safety, a sense of being held and sustained by powers far larger than anything we can comprehend. A safety that comes from being merely an anonymous part of What Is, from feeling oneself as a clutch of sodden soil and hollow bones with the same wind blowing through them that gusts the high ridges. From being in intimate alliance with the bedrock. It's the weird security of realizing that one is a part of something so damned huge.

And so I'm in the grip of this strange, skin-tingling sensation of my own insignificance as I climb the steep slope of a butte, a peninsula of sorts jutting out from a larger escarpment into the desert air. My legs negotiate the gradient by following a crevasse upward until it branches into several smaller ravines. I choose one and follow it to a still smaller tributary, these various creases affording the footholds that finally bring me up near the crest of the scarp. As my eyes rise above that crest, a white shape juts into view, close by, too perfect in its right angles and hence completely incongruous in this place. It takes me two more steps to recognize, startled, that I'm gazing at a metal sign set atop a wooden post. Below and to the right of that white rectangle there's a larger metal cylinder, blocking much of the sign from my low vantage. I climb onto the ridge and these letters emerge from behind what I now see is a metal garbage pail:

PLEASE DEPOSIT
WASTE HERE ⇓

As soon as I see these black words on the white rectangle, the weirdest thing happens. The entire landscape deflates, like a pricked balloon (not with a pop, yet with an almost audible *whoosh*), shrinking abruptly down to about a fourth of the size it just had.

The castled buttes and ridges all around me abruptly forfeit their daunting height. They now seem only slightly taller than where I stand, and that deep, purgatorial valley I've just been stumbling through now appears not so deep after all, indeed a fairly shallow affair. I can easily discern, down over there, some individual shrubs and tumbled rocks I was exploring a while ago, but now somehow they are oddly near at hand—just a hop, skip, and a jump away. This forbidding land, whose immensity had been sending shudders through my spine, now reveals itself as an interesting but very human-scale terrain, and it's almost impossible to believe that I've been wandering lost in this place for much of the day.

I toss into the garbage pail a scrap of paper scavenged while hiking, and follow the asphalt path toward whatever restroom or information booth or parking lot waits to receive me.

❦

There is something eerie about the ability of the written word to shrink the elemental power of a place. Something bizarre about the power of printed letters, even a few painted words on a metal sign, to domesticate the bursting-at-the-seams agency of the wild. It is a uniquely efficacious magic, this alchemic ability of printed words to gather in the expansive *mana* or *wakan* of a place, concentrating it within themselves so that they now speak with the very energy that was a moment ago radiating from the ocher-pink cliffs and shimmering the edge where those ridges meet the indigo sky.

The rock escarpments have now shed their stark vitality. The shadows spreading along those folded cliffs no longer exchange secrets with my muscles, for this other influence grips my attention, drawing me back to my civilized senses.

It is the influence of the verbal mind in ceaseless conversation with itself, dialoguing with its own symbols. The scattered bushes and the shadowed cliffs have no real part in this conversation. They take up their accustomed position, subdued, as the pretty but mostly passive backdrop against which my cogitations now unfold. Given the garbage pail and the paved path, I suppose the highway must be nearby, and so I'll likely be able to hitch a ride, soon, back to wherever my car's waiting. If I'm lucky I'll be able to find a diner still open in some nearby town.

The afternoon's wordless silence, and the edge of danger that had nudged my skin into such an acute and animal alertness, have now dissipated, replaced by the comforting security of a completely human-made world.

Today our relation to the enfolding earth is filtered through a dense panoply of technologies—from air conditioners that mask the heat, to electric lights that hide the night, from capsuled automobiles that hustle us hither and yon to earbuds and headsets whose self-enclosed sounds eclipse the layered silence of the land, blotting out the hum of bees and the whooshing wind whose voice swells and subsides into the belly of that silence. And then there's the ever-expanding array of screens that we daily stare at, summoning images and information up onto their gleaming surface with a few clicks of the keys—this smooth and scintillating surface that's so much more accommodating of our desires than the irregular ground that supports our bodies, so much more responsive to our impatient cravings than the slow reticence whereby seedlings and weeds emerge from the dark depths to become visible against the surface of the soil. Meanwhile, the "nature programming" on television offers such compelling nonstop escapades to our ready gaze—elephants amorously mating, two bull elk locking antlers with a loud clatter, a raptor carrying a hapless young squirrel aloft—that when we walk outside during a commercial we can't help but wonder, seeing how placid things are in our backyard, just where all the action is. The eventful "nature" on the TV seems far more real, somehow, than this slow and rather boring substitute going about its business outside the window.

It is a commonplace to observe that today the perceived world is everywhere filtered and transformed by technology, altered by the countless tools that interpose themselves between our senses and the earthly sensuous. It is less common to suggest that there's a wildness that still reigns underneath all these mediations—that our animal senses, coevolved with the animate landscape, are still tuned to the many-voiced earth. Our creaturely body, shaped in ongoing interaction with the other bodies that compose the biosphere, remains poised and thirsting for contact with otherness. Cocooned in a clutch of technologies, the nervous system that seethes within our skin still thirsts for a relatively unmediated exchange with reality in all its more-than-human multiplicity and weirdness.

Of course, there can be no complete abolishment of mediation, no pure and unadulterated access to the real. Like other social species, we two-leggeds communicate intensely among ourselves, and the various languages we've evolved to do so are themselves a kind of filter that mediates our experience. Still, some ways of speaking are more abstract than others, and some are more permeable to the hooved or scaly shapes that flex and slither through the sensible surroundings. Certainly civilization has many highfalutin forms of speech that hold us aloof from our animal bodies and from the material ground underfoot. And the pervasive digitized forms of discourse that today bring instant communication with persons on the far side of the planet regularly interrupt any remaining rapport between our flesh and the sensuous locale. Yet there are other, older human discourses whose sounds still carry the lilt of the local songbirds, languages whose meanings are less removed from the intimacy of antler and seed and leaf. Such languages live more on the tongue than on the page or screen. They thrive in societies that have not until recently been greatly influenced by the printed word—cultures where human meaning has not yet become wholly ensconced in a static set of visible signs.

While every human language intercedes between the human animal and the animate earth, writing greatly densifies the verbal medium, rendering it more opaque to the many non-human shapes that dwell out beyond all our words. Non-written, oral languages are far more transparent, allowing the things and beings of the world to shine through the skein of terms and to touch us more directly. Since the phrases of an oral culture are not fixed on the page, the sounding of those phrases readily alters from season to season—as the shifting pulse of the crickets may alter the rhythm of our speaking, and even the calm solidity of a boulder we lean against can influence the weight of our spoken words. In such cultures, humans converse less with their written-down signs than with the other speaking powers that stutter and swerve through the soundscape (with the syncopated chanting of toads, and the magpie whose rough soliloquy tumbles down from the upper branches). For here *everything* is expressive, a thunderstorm no less than a hummingbird. To the animistic, oral sensibility, a cedar tree's hushed and whispered phrasings may be as eloquent as a spider's fine-spun patternings, or the collective polyphony of a pack of wolves.

Such elemental, oral languages are rapidly vanishing today, along with the diverse ecosystems that once held them. While linguists estimate that around 6,000 oral languages are currently in use around the world, fully half of them are no longer being taught to children. Which is to say that 3,000 such languages are already effectively lost. Of roughly 175 native languages currently spoken in the United States, 55 are today spoken by fewer than ten people, and only *twenty* are still being spoken by mothers to their young children. It is hardly a coincidence that so many native languages are unraveling at the very same historical moment when innumerable local ecosystems on every continent are fragmenting and falling apart. The conjunction makes evident how deeply the internal coherence of an oral language is entwined with the vitality and coherence of the land itself.

For the peoples who speak such languages are those who steadily engage the land with their muscled limbs and their dreaming senses, those who draw their sustenance—their food and their clothing and the materials for tools and shelter—not from supermarkets or shopping centers but directly from the animate earth. Yet these lifeways, too, are vanishing—as such peoples are displaced from ancestral lands and forced to forfeit traditional livelihoods. Some are indigenous horticulturists, and have been so for centuries, even millennia, like the Mesoamericans who carefully cultivated the earliest forms of maize, or corn, some six thousand years ago. Others are traditional fishermen who work the rivers and the coastal waters with woven nets, or with boats fashioned from hollowed-out trees. For countless generations, such tribal peoples foraged widely for rooted foods, often scattering seeds in particular places as they wandered. Some followed the animals in their migrations, apprenticing themselves to particular creatures. They learned by steady observation how to mimic the calls of those animals and to mime their intricate movements, honoring the creatures in dance and song so those animals might honor them in turn by granting success in the hunt. Others apprenticed themselves to particular plants—smelling, sometimes tasting, slowly ingesting the teachings taught directly by the wild-flowering intelligence of the soil.

The oral languages spoken by such peoples held them close to the speaking earth. As diverse in their lifeways as the contrasting landscapes of the planet, few of these peoples would ever describe or define themselves in isolation from the living land that sustained them.

Thus it is that some of the experiential concepts articulated in this work have less in common with literate styles of reflection than with various states of mind common to indigenous, oral peoples. Although such states may feel peculiar to the modern intellect, it is worth recalling that we all have our indigenous ancestry, and indeed that *our hunter-gatherer heritage is by far the largest part of our human inheritance.* Human culture was itself born in a thoroughly oral context, informed by songs and spoken stories for many tens of thousands of years before any such stories were preserved in a formal writing system. So while the intensely participatory, or animistic, frame of mind common to oral cultures may seem odd to us, it is hardly alien: it is the very form of awareness that shaped all human communication for better than 95 percent of our cultured presence within the biosphere. It is that modality of experience to which the human organism is most closely adapted, the mode of consciousness that has most deeply defined our imagination and our intelligence. We could never have survived, as a species, without our propensity for animistic engagement with every aspect of our earthly habitat. And yet this highly adaptive style of experience has lain mostly dormant in the modern era.[Ω]

❧

While indigenous, oral cultures tend to be exceedingly different from one another, it is nonetheless possible to discern several intu-

[Ω] There are many intellectuals today who feel that *any* respectful reference to indigenous beliefs smacks of romanticism, and a kind of backward-looking nostalgia. Oddly, these same persons often have no problem "looking backward" toward ancient Rome or ancient Greece for philosophical insight and guidance in the present day. What upsets these self-styled "defenders of civilization" is the implication that civilization might have something to learn from cultures that operate according to an entirely different set of assumptions, cultures that stand *outside of* historical time and the thrust of progress. Many persons steeped in Western science tend to assume that native notions are superstitious or simply naïve, unaware that indigenous thought stems from a radically different view of what language is, and what thinking is for.

There is simply no way to comprehend indigenous notions without stepping aside from commercial assumptions that are broadly taken for granted today (including the basic equation of *land* with *property*—with a commodity that can be bought, sold, or owned). Indigenous insights cannot be understood without slowing down, without taking time to notice the upward press of the ground and the earthen silence that surrounds all our words. Often at home in such silence, oral peoples tend toward reticence, reluctant to broadcast their experience very loudly. Hence, while indigenous traditions are vigorously unfolding today, the philosophical intensity and practical wisdom of native peoples all too often remains invisible and unheard amid the bustle and blare of contemporary commerce, conveniently ignored by those who most have need of such intelligence.

itions, and qualities, that they hold in common. Such generalizations are always problematic, and prone to be misread. But if we rigorously remember and affirm that any such common intuitions necessarily express themselves differently—*very differently*—for each specific people, then perhaps we may dare to list a few of the qualities endemic to the oral frame of mind:

FIRST: Oral awareness is intensely local in its orientation. Without the many communication technologies recently spawned by the printed word (without the ability such media have to bring us into contact with far-off places), indigenous, oral awareness is much more deeply informed by the immediately surrounding locale than most modern folks can even imagine. In the absence of intervening technologies, our senses—coevolved for millions of years with the textures, colors, and sounds of surrounding nature—spontaneously couple themselves to shapes and dynamic patterns in the living landscape, tracking those patterns as they metamorphose through the seasons. The leafing time of the local trees, the rhythms of bud and blossom and fruit, the reticence of various animals in certain months and their antic exuberance in others—all these unfoldings in the immediate environs provide a set of sensuous metaphors for the complex pulse of our own emotions, and a basic template for our cogitations. The human animal is a creature of imagination, to be sure, yet our imagination is first provoked and infused by the earthly *place* where we dwell, or by the wider terrain wherein we circulate. Indigenous, oral intelligence is place-based intelligence, an awareness infused by the local terrain.

SECOND: The simple act of perception is experienced as an interchange between oneself and that which one perceives—as a meeting, a participation, a communion between beings. For each thing that we sense is assumed to be sensitive in its own right, able to feel and respond to the beings around it, and to us.

THIRD: Each perceived presence is felt to have its own dynamism, its own pulse, its own active agency in the world. Each phenomenon has the ability to affect and influence the space around it, and the other beings in its vicinity. Every perceived thing, in other words, is felt to be animate—to be (at least potentially) alive. Death itself is more a transformation than a state; a dying organism becomes part of the wider life that surrounds it, as the hollowed-out trunk of a fallen tree feeds back into the broader metabolism of the forest. There is thus no clear divide between that which is animate and that which is inanimate. Rather, to the oral awareness, *everything* is animate, *everything* moves. It's just that some things (like granite boulders) move much slower than other things (like crows or crickets). There are only these different speeds and styles of movement, these divergent rhythms and rates of pulsation, these many different ways of being alive.

The surrounding world, then, is experienced less as a collection of objects than as a community of active agents, or subjects. Indeed, every human community would seem to be nested within a wider, more-than-human community of beings.

FOURTH: The ability of each thing or entity to influence the space around it may be viewed as the expressive power of that being. All things, in this sense, are potentially expressive; all things have the power of speech. Most, of course, do not speak in words. But this is also true of ourselves: our own verbal eloquence is but one form of human expression among many others. For our body, in its silence, is already expressive. The body, itself, speaks.

FIFTH: Since our own sensitive and sensuous bodies are entirely a part of the world that we perceive—since we are carnally embedded within the sensuous field—then we can experience things only from our own limited angle and place among them. Hence we have only a partial view of each entity or situation that we encounter; there is no aspect of this world that can be fathomed or figured out by us in its entirety. There are always aspects that are hidden from view, dimensions that we cannot perceive directly. The depth of the world, and indeed of any part of the world, is therefore inexhaustible. Every certainty, every instance of clear knowledge, is necessarily surrounded by a horizon of uncertainty shading into mystery.

SIXTH: To an oral culture, the world is articulated as story. The surrounding cosmos is not experienced as a set of fixed and finished facts, but as a story in which we (along with the moon sliding in and out of the clouds, and the trout leaping for a fly) are all participant.

For the relation of a tale to its characters is much the same as the relation of this earthly cosmos to its inhabitants. Just as there is an interiority to the perceived world (carnally enfolded as we are by the round expanse of the terrain and the curving vault of the sky), so the characters in a well-told tale live and breathe within the voluminous interiority of the story itself.

In other words, we find ourselves situated in the land, with its transformations and cycles of change, much as protagonists are situated in a story. To a deeply oral culture, the earthly world is felt as a vast, ever-unfolding Story in which we—along with the other animals, plants, and landforms—are all characters.

SEVENTH: In such a breathing cosmos, time is not a rectilinear movement from a distant past to a wholly different future. Rather, time has an enveloping roundness, like the encircling horizon. It is a mystery marked by the slide of the sun into the ground every evening and its rebirth every dawn, by the incessant cycling of the moon and the round dance of the seasons. The curvature of time is here inseparable from the apparent curvature of space; and indeed both remain rooted in the round primacy of place. For each place has its particular pulse. Each realm has its rhythms, its unique ways of sprouting and unfurling and giving birth to itself again and again—as the world itself turns and returns, and as indeed the best stories are told over and over again.

EIGHTH: A world made of story is an earth permeated by dreams, a terrain filled with imagination. Yet this is not so much *our* imagination, but rather *the world's* imagination, in which our are participant. As players within an expansive, ever-unfolding story, our lives are embedded within a psyche that is not primarily ours.

The dreamy, emotional atmosphere that permeates a story is much like the fluid atmosphere that enfolds our breathing bodies, with its storms and its calms. Awareness itself is here inseparable from the air—from this invisible medium, infused with sunlight, which circulates both within and all around us, binding our life together with that of the tempest and the swaying pines . . .

So mind is not experienced as an exclusively human property, much less as a private possession that resides within one's head. While there may indeed be an interior quality to the mind, for a deeply oral culture this interiority derives not from a belief that the mind is located within us, but from a felt sense that *we* are located *within it,* carnally immersed in an awareness that is not ours, but is rather the Eairth's.

NINTH: Each entity participates in this enveloping awareness from its own angle and orientation, according to the proclivities of its own flesh. We inhale the awakened atmosphere through our skin or our flaring nostrils or the stomata in our leaves, circulating it within ourselves, lending something of our unique chemistry to the collective medium as we exhale, each of us thus animated by the wider intelligence even as we tweak and transform that intelligence. The rooted beings among us twist and flex in the invisible surge; other creatures are carried aloft by the whirling currents. The denser life of rock may seem impervious to those winds, yet the crevassed contours of the mountains have been carved over eons by the creativity of wind and weather, as those mountains now carve the wind in turn, coaxing spores out of the breeze and conjuring clouds out of the fathomless blue.

The wild mind of the planet blows through us all, ensconced as we are in the depths of this elusive medium. However, although it is our common element, every one of us experiences it differently. No two bodies or beings ever inhabit this big awareness from precisely the same angle, or with the same sensory organization and style. Since the body is precisely our interface and exchange with the field of awareness, a praying mantis's experience of mind is as weirdly different from mine as its spindly body is different from mine; and the dreaming of an aspen grove is as different from both mine and the mantis's as its own fleshly interchange with the medium is different from ours. It is our *bodies* that participate in awareness. Hence no one can feel, much less know, precisely how the big mystery reveals itself to another.

Here's another way this might be said: each of us by our actions is composing our part of the story in concert with the other bodies or beings around us. Yet since we are situated *within* the story, dreaming our way through its voluminous depths according to the unique ways of our flesh, no one of us can discern precisely how the story can best be articulated by another. No human individual can fathom just how the encompassing imagination is experienced by any other person—much less by a turtle, or a thundercloud, or by a car door patiently rusting at the junkyard, its viridian paint flaking off in the desert heat.

Our carnal immersion in the depths of the Mysterious thus ensures an inherent and inescapable pluralism. And yet—and yet: although there is no single way to tell it, it is the same Tale that is unfurling itself through our gazillion and one gestures. It remains the same Eairth whose life-giving breath we all inhabit, the very same mystery that we each experience from our own place within its depths.

You may be dismayed, kind reader, when reading that last point just above—flummoxed by my use of so many different terms to speak of the common earthly sentience in which our various sensitive bodies are situated. I write of it as a huge "imagination," as "psyche" and as "mind," as the breath of the planet, as an encompassing story, the big "dreaming" in which we're corporeally immersed. But please recognize this: that I vary my terms entirely on purpose, in order to indicate that it is not a mere word that I speak of, but the enigmatic experience toward which these words point, a palpable but unseen medium that lies beyond all our concepts.

For it IS palpable, a dimension as tangible as the rushing air, although like the air it can neither be grasped by our hands nor rendered visible to our eyes. But again, you protest: Am I not pushing this simile much too far? By likening the psyche, or mind, or indeed awareness itself to the encompassing air (to this taken-for-granted element that is even now sliding in and out of your throat), do I not risk falsifying the psyche, rendering a manifestly immaterial power as a mundane material substance, inducing people to take literally what is merely an extravagant metaphor?

I think not. For there is something much more than a metaphor here—an ancient and elemental kinship between air and awareness, between the mind and the wind. It is a kinship demonstrated, as we have noted, by the Aeolian etymology of such words as "psyche," "spirit," and "anima" (itself the Latin word for soul, derived from the ancient Greek *anemos,* meaning "wind"), and by the Indo-European etymology of the word "atmosphere" (which shares the same origin as the Sanskrit word *atman,* meaning "soul"). In truth the modern, civilized understanding of *mind* as a purely immaterial power has been born by a process of subtraction, slowly and by increments, from the ancestral experience of the invisible atmosphere as a thick, meaning-filled plenum in which we're immersed—as a living field of intelligence in which we participate. To our oral, indigenous ancestors, the animating air was the very place of the spirits, the very medium of awareness.

Such a sensuous articulation of mind is very far from the conventional way of saying things in the modern and postmodern world. Indeed, *every one* of the nine qualities enumerated above is alien to the abstract style of thought common to our broadly literate, technologically informed, steadily globalizing civilization. Yet the chapter-by-chapter investigations in this book—the patient analyses of depth, shadow, house, gravity, and ground as experienced by our animal bodies, the careful questioning of the felt meanings conveyed in birdsong and the gestures of boulders, the descriptive investigations of shapeshifting and interspecies rapport—have shown that many of the oral traits just enumerated remain active just below the surface of our civilized consciousness. The many explorations of sensory experience in this work—the experiential studies of mind and mood and weather, disclosing ways in which the human organism tacitly dialogues with its changing surroundings—all indicate that such oral, animistic proclivities remain fiercely operative within even our most taken-for-granted encounters with the world.

To the extent that conventional discourse hides this primary dimension of experience (to the extent that our everyday ways of speaking steadily *deny* such felt participations), our tongues enact a massive split between our minds and our bodies, effectively severing our verbal, speaking selves from our corporeal, animal experience.

Of course, the modern, technologically literate self—sophisticated, cosmopolitan, abstractly interested in almost anything yet coolly detached from earthly reality—is not the only possible constitution of the self. Oral, storytelling cultures, with their heterogeneous, place-based economies rooted in barter and gift exchange, engender a very different kind of self than literate culture, with its cosmopolitan market economy, or than digital culture, with its homogenizing global monetary systems. Indigenous, oral culture not only constitutes the original and longest-standing form of human society, it also gives rise to, and remains correlative with, the most elemental layer of the human self or subject.

The oral self, in other words, was the first formation of the human subject as a speaking being, as a creature of culture. Archaeological and ethnological evidence from every inhabited continent attests that this primordial form of the cultured human animal was not an abstract or intellectual self, but an embodied, viscerally felt identity informed by, and entangled with, the perceptual surroundings. Communal, deeply animistic, yet keenly awake to the elemental practicalities of water and food and shelter, our ancestral and somewhat collective sense of self remained inseparable from both the body and the terrain. For the human body was here experienced not as a determinate object but as a volatile and metamorphic presence, sometimes concentrated and sometimes dispersed, a conjunction of material powers slipping in and out of itself with the breath. Intertwined with the other bodies that surround—not only with other persons but also with other animals and plants and landforms—this corporeal self was hardly mute. It was, rather, storied: a tangible yet shifting identity that expressed itself in narrative and song, in spontaneous, rhythmic chants (composed often of *vocables*—expressive sounds without determinate meaning), and in whispered prayers offered on the breath to capricious powers in the enfolding field.

Something of this ancestral self stirs and awakens whenever we tell, with the whole of our body, a story that's rooted in the local land, or whenever we listen to such a tale being told, not by some personage on a screen or a disembodied voice on the radio, but by a palpable person who stands before us, inhaling the same air that fills our own lungs as she gestures toward the gleaming crescent rising like an elephant's tusk above the trees. It stirs in us when we wander down to sit with our child on the stones by the river, dangling a line below the surface while recounting a tale, first heard from our grandfather, about a talking fish who once granted him a simple wish in return for its watery life. Or here, in the high desert, when one comes upon a ropelike bit of scat, woven from crunched bones and mouse fur, judiciously placed like an indignant sign right in the middle of the path. For the ready recognition of that sign's author sparks the memory of one or another story wherein that same Coyote, like a holy fool, accidentally upends the world.

But tricksters come in many forms. There are certain late nights, driving through the empty streets of the nearby city, when every green light turns to red just as I approach the intersection. Upon skidding to a stop at the sixth or seventh such insult, I suddenly recall how—in so many of the old tales—the spirits tend to linger and congregate *at the crossroads.* Aha! So those spirits are still here, drawn by the electric hum of the wires, flipping the traffic lights to red whenever they sense my approach, feeding gaily on the sparking rage of my resultant frustration: I'll use the bicycle next time.

Whether in the heart of the city or the thick of the wilderness, our indigenous soul stirs and comes awake whenever we find ourselves thinking in storied form, and so the buildings lean toward us and the trees in the backyard begin to speak in low, groaning tones as the trunks rub against one another. If we are thinking in literate, logical terms then these tones are not voices, but when we're thinking in stories then they are indeed a kind of speaking, for to the oral imagination every entity has its eloquence, and so our muscled flesh can't avoid the sense that those sounds are filled with expressive meaning, as even the few clouds and the clustered rocks are alive with felt meanings, though we can hardly translate that meaning into words. The breeze is an elixir carrying the green chemistry of the needles up through the double arch of our nostrils to burst as a steady tang on the moist membranes inside, while the autumn blue of the sky, as it filters through the branches, is itself a kind of wine casting a giddy charm upon our limbs, making us crouch and leap with pleasure. This whole terrain is talking to our animal body; our actions are the steady reply.

That such participatory experiences remain accessible for many of us even in the midst of the technologized world—that they have not been eradicated by our more sophisticated ways of seeing and thinking—indicates that here there is something basic to the very constitution of the human creature, something necessary to our ongoing vitality as a species. That such animistic inclinations remain active underneath all our literate logics does not invalidate those more recent and refined logics—not at all! But it does suggest that our abstract forms of reflection remain dependent, in some manner, upon this older and more full-bodied mode of experience. The stubborn persistence of participation suggests that this ancestral form of experience is the hidden ground in which reason remains rooted, the secret soil from whence all later forms of reflection draw their sustenance.

The primordial impulse to animate and participate with our terrestrial surroundings has long been channeled by sedentary civilization into a more focused participation with a Source presumed to dwell beyond the perceivable cosmos. We have already seen how formal religion, by concentrating our age-old animist proclivities upon a single, omnipotent agency located *outside* the apparent world, loosened our respectful relation to the sensuous surroundings. The religions of the Book freed civilization to manipulate those surroundings with a more utilitarian abandon, unleashing a steadily rising tide of invention and innovation. With our relational yearnings focused on a transcendent power now hidden entirely beyond the natural world, we were free to delve more boldly into the depths of earthly nature without heed of apparent impropriety, following our curiosity into every nook and cranny, ultimately extending our manipulations into the nucleus of the atom, and into the nucleus of the living cell.

Yet our impulse toward participation, our yearning for engagement with a more-than-human otherness, has never been eradicated. The mass turn toward formal religion by citizens of industrialized nations in the last half century, the swelling membership in fundamentalist creeds of every kind, the myriad new shapes of belief that adorn the spiritual marketplace, all give evidence that the human craving for relation with that which *exceeds* us is as strong as ever. Nonetheless, given the ancestral impulse toward otherness at the core of this craving, it is not likely to be finally sated by any God made in our image, any more than it can be satisfied by our human-made technologies. A much older and deeper accord is at stake here, an alliance as intimate as the breath, and one as easily overlooked by the intellect.

When we speak of the human animal's spontaneous interchange with the animate landscape, we acknowledge a felt relation to the mysterious that was active long before any formal or priestly religions. The instinctive rapport with an enigmatic cosmos at once both nourishing and dangerous lies at the ancient heart of all that we have come to call "the sacred." Temporarily forgotten, paved over yet never eradicated, this old reciprocity with the breathing earth was here long before all our formal religions, and it will likely outlast all our formal religions. For it has always been operative *underneath* our various religions, nourishing them from below like a subterranean river.

There is no disdain for religion in such a statement. We can honor the awesome eloquence of each religion while acknowledging the precarious nature of church-based faiths in today's crowded and crisis-ridden world, where people of divergent scriptures must somehow learn to get along. Our greatest hope for the future rests not in the triumph of any single set of beliefs, but in the acknowledgment of a felt mystery that underlies all our doctrines. It rests in the remembering of that corporeal faith that flows underneath all mere beliefs: the human body's implicit faith in the steady sustenance of the air and the renewal of light every dawn, its faith in mountains and rivers and the enduring support of the ground, in the silent germination of seeds and the cyclical return of the salmon. There are no priests needed in such a faith, no intermediaries or experts necessary to effect our contact with the sacred, since—carnally immersed as we are in the thick of this breathing planet—we each have our own intimate access to the big mystery.

Each of us must finally enact this rapport in our own unique manner, discerning and learning to trust the particular gifts of our flesh even as we draw insight from the ways of others. Slowly we come to follow the promptings of our heart as it responds to the larger pulse of this earthly cosmos, listening inward even as we listen outward. And thus our voice, tentative at first, finds its own improvisational place in the broader polyphony—informed by, yet subtly altering, the texture of that wider music. Our rapport is ours alone, and yet the quality of our listening, and the depth of our response, can transform the collective texture of the real.

While exploring, some years ago, the bustling streets of a large Canadian city, I passed by some store windows displaying an unusually broad array of sacred texts translated from the world's great spiritual traditions. I opened the door and stepped inside. The ribbon of incense unfurling in the air indicated that I had entered one of the new-age bookstores that sprang up in various cities during the latter half of the twentieth century in response to a deepening thirst for things spiritual. This store was far larger than any I'd yet encountered. Tibetan thangkas hung on the walls of the several rooms, and plush meditation cushions were scattered about, while the accumulated wisdom of all the ages seemed to be resting upon the crowded shelves. Entire aisles were dedicated to different Christian traditions and sects (from the Gnostics to the Quakers), each with their respective mystics. Another corridor was lined with Islamic and Sufi tracts in translation (including multiple shelves filled with the poetry of Rumi and other ecstatics), and still another overflowed with books on the Jewish mystical tradition or Kabbalah (translations of the Sefer Yetzirah and the Zohar, compilations of Hasidic folklore, various contemporary neo-Hasidic texts). Each of the primary Buddhist traditions had its own aisle—Theravada Buddhism (and Vipassana meditation); Vajrayana or Tibetan Buddhism (including translations of the Tibetan Book of the Dead, and many tracts written by or about the current Dalai Lama); and Zen (with its abundant lineages and contemporary exemplars). In the smaller Taoism section, contrasting translations of the Tao Te Ching jostled with one another (selflessly) for space.

Hindu teachings took up another aisle and a half; I saw translations of the Ramayana and the Mahabharata and numerous commentaries on the part of the latter called the Bhagavad Gita, large tomes on Vedantic philosophy, as well as many small, self-published books containing the sayings of particular yogins. Celtic folklore took up many shelves. Diverse and divergent Native American traditions had their own corner of the store, including creation stories and folklore from different tribes, along with anthropological studies, books by indigenous activists, and the transcribed teachings of revered elders and medicine persons from many parts of the Americas. Several shelves were devoted to Aboriginal Australian art and mythology, and a large aisle was given over to the great African mythologies and cosmologies both on the continent and in diaspora.

Wandering from room to room among all these colorful and carefully ordered tomes soon made me dizzy, overwhelmed by the discovery that such an abundance of wisdom could be gathered in one place. I came to a halt, finally, in the middle of the largest room, trying to take it all in. Here were the deepest insights of humankind, instructions of incalculable value garnered from every era and every region of the earth. I simply had not realized that the sacred teachings from many of these esoteric traditions were even open to the gaze of outsiders—much less that they had been translated into various contemporary languages, and so rendered accessible and available to so many contemporary persons. But it was not this accessibility, alone, that stunned me.

It was also the strange incongruity of realizing that, while such an inexhaustible plenitude of ethical intelligence was today open and available (was being read, integrated, and disseminated within contemporary culture, was being taught in college classrooms while also being pored over and pondered by countless private individuals), still, the heedless desecration and ruination of the more-than-human natural world was accelerating all around us. The last of the great forests were being laid waste, more and more wetlands were being paved over, oceanic dead zones (born of the chemical runoff from our industries) were expanding and multiplying across the globe. Greater numbers of our fellow creatures were losing their habitual haunts every year, their migratory routes severed by new highways and subdivisions, their bodies choking on our toxins and contorted by never-before-seen climatic conditions; more and more species were being squeezed out of existence by our current ways of life.

How to make sense of this juxtaposition? How could the open abundance of moral wisdom that was now accessible to any and all seekers within contemporary society be reconciled with the utter *callousness* of that same society, with the steady collusion of its citizens in the destruction of so much wonder? How could a culture as educated and spiritually curious as ours—a culture exemplified by the persons around me drawing assorted volumes from the shelves, cruising their pages with raised or furrowed eyebrows while standing in the aisles—how could such a culture be so oblivious, so reckless, in its relation to the animate earth?

I turned and walked, once again, past bookcases thick with teaching stories from indigenous tribes of the Americas, touching the spines of various volumes, absentmindedly sliding out a text on the animal folklore of the Pueblo Indians. How remarkable, I thought, to be able to find teachings so specific to the southwest desert way up here in a Canadian bookstore . . .

And then it hit me, clear as the whack of a wooden mallet on a bronze meditation gong. No wonder! No wonder that our sophisticated civilization, brimming with the accumulated knowledge of so many traditions, continues to flatten and dismember every part of the breathing earth. No wonder that, despite all we may have learned of the ethical intelligence of the ages, we remain so oblivious, so impervious, to the rest of nature!

For we have written all of these wisdoms down on the page, effectively divorcing these many teachings from the living land that once held and embodied these teachings.

Once inscribed on the page, all this wisdom seemed to have an exclusively human provenance. Illumination once offered by the moon's dance in and out of the clouds, or by the dazzle of sunlight on the wind-rippled surface of a mountain tarn, was now set down in an unchanging form. Guidance that once came from the complex interplay of elemental forces in the dank heat of the rain forest, visionary insights that arose among peoples hunkered in the endless twilight of an Arctic winter, could now be carried elsewhere, read in distant towns or on distant continents by readers at far remove from the actual textures and tastes that once informed all these insights. And so the place-specific intelligence that originally infused all these many teachings would be forgotten. By writing all this relational knowledge down on the page, we tore these teachings from the actual earth that once taught them to us, detaching them from the particular climates and seasons that first provoked such insights. We severed all this intelligence from the dense mountain forests and the migrations of game, from the frozen winters and the sweltering deserts that had forced our lives into new patterns, from the watering holes thick with animals where these dangerous insights swam up from the depths or alighted on our shoulders or otherwise revealed themselves to us and became a part of our knowing.

So many of the teaching tales in these books—like the sacred stories of the Haida or the Iroquois or the Lakota, like the holy legends of Tibet or the wisdom tales of Ireland—were once embodied in particular landscapes, in the meandering abundance of a particular river and the uncanny shape of that mountain crag, in the healing medicine of this sacred spring bubbling out of the ground. To glimpse the mouth of a sheltering cave in the distance, or to reach the edge of a muddy lake, was to remember the mythic story about how that cave came to be there, and which old ancestor's blood, gushing from a fatal wound, formed the ruddy color of that lake. *The local earth, in effect, was the primary mnemonic, or memory trigger, for remembering the oral tales.*

The stories were carried, as well, by the seasonal nourishments and poisons of particular plants—by the stamina afforded by certain well-chewed leaves, and the transformative power held by certain mushrooms, or the curative medicine of specific strangely shaped roots. Many other stories held within them the dangerous grace of particular predators—some of whom, it was told, had married into our human lineages.

A young woman, out picking berries, comes upon a mound of bear shit; rather than detour around the trace of this powerful being, she playfully leaps back and forth over the pile, and kicks some of the dried dung along the path as she walks. Soon she's lost. A strange man, large and handsome, approaches from out of the woods, offering to help her find her way. Following him back to his solitary camp, falling in love, only much later does she realize that she's wedded herself to a large grizzly; even as she notices the new fur growing on her arms, she's now pregnant with their cubs—half human and half bear.

Similarly, a young fisherman finds and hides the dark, oily cloak of a naked woman he spies sleeping on the beach. Beautiful and taciturn (and unable to find her clothes), she comes home with him, and soon becomes pregnant with his children. Years later, when she discovers her old, forgotten cloak and disappears into the waves, a recognition dawns within the fisherman that the woman he married was really a seal; he now knows that their dark-eyed children will be intermediaries between the human world and the finned peoples of the sea.

Other stories drew their power from the feathered vitality of the wren, and from the glistening wisdom of the salmon. Indeed, many tales were known to be under the influence of the winged or scaled or antlered beings that moved within them, and of the particular places where those storied events once happened and might happen again. And hence those stories were told or sung only in the shadow of those cliffs, or when setting the fishing nets, or to honor the wild spirit of Old Honey-Paws immediately after a successful bear hunt.

Various teaching tales were acted out during the migrations of certain animals, while others were told only when harvesting specific herbs. Each valley had its stories, which were echoed in the diverse place-names carried by the boulders and the meadows and the riverbends of that realm. The stories, in turn, had their particular seasons for the telling. For the living land was felt to be the primary author of these tales! The earthly cosmos, flowing its life-giving waters through different regions, was the primary mystery articulating itself through these many teachings. Human life was obviously a part of this wider life; if the land fell sick, if other animals were vanishing or the forest trees began dying, it was obvious that the human community would soon falter as well.

By writing oral traditions down, we thought simply to preserve them, and to render their teachings more accessible. We did not realize that in order to plant them on the page we were uprooting these deep teachings from the soils that gave them their specific vitality. We didn't suspect that by transcribing them on the page we were stealing away the expressive power of each place, usurping the manifold eloquence of the land and translating it into a purely human tongue. We didn't realize that we were divesting the ground of its voice.

Transcribing previously oral teachings did not *force* the bards and the griots to stop recounting these tales in relation to the animate earth, did not *compel* the "wicasa wakan" to cease instructing us in the specific medicines that flourish in particular places. The recording of traditional knowledge in printed books, however, suggested that such face-to-face tellings out in the land were no longer entirely necessary, that the gesturing body of the storyteller—like the palpable presence of those mountains or the tumbling exuberance of the river—was no longer really needed for the remembrance of the storied knowledge, or the preservation of the culture. Indeed, the arrival of writing and written books seemed to render the storytellers superfluous, and the land as well. Now the paper leaves of the book, rather than the chattering leaves of oak and beech and birch, seemed to hold the ancestral knowledge. Slowly the landscape fell mute, as though the many powers lurking within it simply withdrew from human contact, receding back into the heartwood and the density of the bedrock.

So the spread of literacy brought a new detachment from the elemental immediacy of the local earth. We no longer needed to *see* those actual cliffs, coastlines, or creekbeds in order to recall the countless stories (and the layered information stored in those stories). The book, rather than the sensible landscape, became the primary mnemonic for remembering the ancient tales. And again: once written down, the originally oral tales became far more portable. Although I've here been describing some of the losses that this entailed, it's time to acknowledge that this portability is also one of the greatest *gifts* of the written word. Books have enabled us to partake of stories from vastly different places, ingesting insights originally rooted in other traditions and other eras. By exposing us to ideas drawn from many divergent cultures, literate society has in truth brought an abundance of blessings—instilling what is often a useful distance, and detachment, from our immediate environment. *The culture of the book,* we might say, *is inherently cosmopolitan.*

In our own era, a very new kind of literacy has been conjured forth by the advent of the personal computer. In the prosperous parts of the world, the digital literacy of the World Wide Web already provokes far greater engagement than the sphere of books, newspapers, and magazines. Our involvement with the Internet—with e-mail and blogs and social-networking websites—brings us almost instantaneous information from around the planet, empowering virtual interactions with persons and groups in other places.

This new, digital dimension of human interchange is much less embodied than even that of the printed page. Books are relatively easy for our bodies to relate to. Quiescent denizens of the so-called analog world that we carnally inhabit, books can be hefted and smelled, their pages bent or bitten and tasted. But the digital letters and images dancing across the screen are envoys of a far more abstract technology—luminous phantoms, effervescences at the surface of a vast and ever-shifting ocean of largely disembodied data. The person composing a blog on her laptop commonly knows very little about how all that hardware and software functions; she simply trusts herself to the technology. Synapsing herself to the terminal, she navigates as a weightless mind in the fathomless sea of information, sometimes interacting with other bodiless minds that have logged on in other locations around the globe. If literate culture is inherently cosmopolitan, *digital culture,* it would seem, *is inherently global and globalizing.*

There is an immense amount to value about both of these more abstract and mediated forms of communication, commerce, and culture. The literary exchange of written stories has enabled the heady cosmopolitan buzz of our cities, and has enlivened our respect for traditions curiously different from our own. Good books deepen our interiority, and they complexify our individuality. Who *I* am, even to myself, has been informed and amplified by the many other selves—from Huck Finn to Hans Castorp, from Anna Karenina to Mr. Palomar and Burley Coulter—who first entered and took possession of me through the medium of the printed page.

The emergence of the Internet, meanwhile, offers a remarkable new autonomy to the individual (providing easy, personalized access to much of the world's information) even as it blurs the apparent boundaries of the self, entwining its users into a complex and ever-ramifying amalgam of circuitry and gregarious sociality. Just as important, it empowers rapid affiliations and alliances between persons who may never meet, enabling activists to mobilize on a moment's notice to block disastrous developments, or to catalyze necessary changes.

Such gifts are useful and worthy adjuncts to our more full-bodied participation in the ongoing life of our community. Yet these lucid gifts can hardly *substitute* for that community. They cannot supplant the necessary nourishment of real, face-to-face engagements with other persons, and other beings, in the actual terrain where we find ourselves. The abundant blessings of the book, like the more recent, pragmatic boons of the computer and the handheld screen, transform into poisons when they occupy the bulk of our waking attention—when our ceaseless interchange with the printed page or the digital screen short-circuits the old, instinctive reciprocity between our senses and the sensuous earth.

Oral culture—the face-to-face sharing of living stories that are *not* written down—is the form of human society that goes hand in hand with such reciprocity. If digital culture is inherently globalizing, and if the culture of the book is inherently cosmopolitan, *oral culture is inherently local in its orientation.* Since the tales that compose a deeply oral culture are not written down, they do not easily travel far from the broad bioregion where they originate, or from the particular landforms and creatures that figure as prominent powers in the tales. There exists a palpable intimacy between language and the land in any deeply oral culture, an eros that binds both together and feeds the vitality of each. For the surrounding earth is aquiver with stories that seem to sprout from every field and forested ridge! The spoken tongue carries the textures and tones of the local topography; the rhythms of the language, in turn, are preserved in the living landscape.

What should be obvious, now, is this: *the global culture of the Internet and the cosmopolitan culture of the book both depend, for their integrity, upon the place-based conviviality of a thriving oral culture.*

Oral, storytelling culture—and the vernacular intimacy with the local land that coincides with such culture—is the forgotten ground that still supports those more abstract layers of culture. It is the neglected but necessary soil from whence civilization still draws its sustenance, the nourishing humus in which our humanity remains rooted.

When oral culture degrades, the mediated mind loses its bearings, forgetting its ongoing debt to the body and the breathing earth. Left to itself the literate intellect, adrift in the play of signs, comes to view nature as a sign, or a complex of signs. It forgets that the land is not first and foremost an arcane text to be read, but a community of living, speaking beings to whom we are beholden. Adept at representing the world verbally, the literate intellect forgets how to orient in the midst of the world's presence, how to hear those many voices that do not speak in words.

Similarly, the computerized mind, when left to its own devices, all too easily overlooks the solid things of the earth. Skilled in the rapid manipulation of symbols, it neglects the stones and the grasses that symbolize nothing other than themselves. Dazzled by its own virtual creations, the digital self forgets its dependence upon a world that it did not create, overlooking its carnal emplacement in the very world that *created it.*

When oral stories are no longer being told in the woods, or along the banks of gushing streams—when the land is no longer being honored *ALOUD* as an animate, expressive power—then the human senses lose their attunement to the more-than-human terrain. Fewer and fewer people are able to feel the particular pulse of their place; many no longer notice, much less respond to, the fluent articulations of the land. Increasingly blind, increasingly deaf—increasingly impervious to the sensuous world—the technological mind progressively lays waste to the animate earth.

Today our vaunted civilization pours its by-products into the winds and into the waters. The weather tilts toward catastrophe; ice caps melt; the snowpack evaporates; water fit for human consumption hides in smaller and smaller oases within the desert of the real. Ever more creatures wane and vanish from this non-virtual reality, unable to adapt to the wrenching changes we've wrought. Massive animals and small animals, hoofed ones and clawed ones, antlered and quilled and bright-feathered ones, finned and tentacled and barnacled ones, all steadily dwindling down to a few members before they dissolve entirely into the fever dreams of memory.

How, then, to renew our visceral experience of a world that exceeds us—of a world that is wider than ourselves and our own creations? Does a revitalization of oral culture mean that we must renounce reading and writing? Must we empty our bookcases? Must we unplug our computers and drag them down to the dump?

Hardly. The renewal of oral culture entails no renunciation of books, and no rejection of technology. It entails only that we leave abundant space in our days for an interchange with one another and with our surroundings that is not mediated by technology: neither by television nor the cell phone, neither by the handheld computer nor the GPS satellite (nor any of the newer digital allurements that promise to arrive in the coming years). Nor even the printed page.

Among writers, for example, it entails a recognition (even an anticipation) that there are certain stories we may stumble upon that ought not to be written down—stories that we might instead begin to tell with our tongue in the particular topography where those stories live! Among parents it requires that we set aside, now and then, the books that we read to our children in order to recount a vital story with the whole of our gesturing body—or better yet, that we draw our kids out of doors in order to improvise a tale about how the nearby river feels when the fish return to its waters, or about the wild wind that's even now blustering its way through the city streets, plucking the hats off people's heads . . .

Among educators, it requires that we begin to rejuvenate the arts of telling, and of listening, in relation to the geographic place where our lessons actually happen. For too long we have incarcerated the potent magic of linguistic meaning within an exclusively human space of signs. Hence an American youth may attend a high school in New England or California, or perhaps a small boarding school at the edge of the Rockies—yet this will make little difference, since she'll be taught largely the same things in each location. Since truth has come, over the centuries, to reside on the printed page, knowledge (it now seems) floats entirely free of place. Even if we research our facts via the latest search engine, surfing from webpage to webpage as we assemble our assignment, still this knowledge has little to do with the animate terrain hollering outside the window. The wildness of where we are remains muzzled, the local earth still mostly a mute and passive backdrop against which human happenings unfold.

Can we renew in ourselves an implicit sense of the land's meaning, of its own many-voiced eloquence? Not without renewing the sensory craft of listening, and the sensuous art of storytelling. Can we help our students to carefully translate the quantified abstractions of science into the qualitative language of direct experience, so that those necessary insights begin to come alive in their felt encounters with cumulus clouds and bleaching corals, with owls and deformed dragonflies and the intricate tangle of mycelial mats? So that the potent evidences steadily emerging from the sciences are no longer employed mostly by profiteering corporations for whom the land is strictly a set of numbers, but by people, young and old, mobilizing to halt such reckless developments? Most important: Can we begin to restore the health and integrity of the local earth? Not without *restorying* the local earth.

The replenishment of oral culture would thus bring a new realization of the primacy of place and proximity, a recognition that genuine community is not, first and foremost, something we create online with those who share our specific values, but something that must be practiced on the ground with our actual neighbors. This would be a tall order for many modern persons, given the contemporary passion for insularity, for tall fences and electronic gates. And yet the payoff would be a dramatic increase in real security. For as is already becoming evident in North America, neither insurance companies nor federal disaster-relief agencies are up to the task of providing relief for a steadily rising tide of climate-related catastrophes. Our most dependable insurance against calamity resides in the fellow feeling and generosity of those who live near us, in the impulse to look out for each other's children or to look in on an elderly neighbor when the electricity's been out for a day. Our truest security lies where it always has, in community solidarity and neighborliness.

Regenerating oral culture would entail beginning to resuscitate local festivals, organizing celebrations to honor the cyclical return of the salmon in dance and song and spoken story, festivals to honor the becoming-fruitful of the trees and the long sleep of the sun at the winter solstice. Telling the tale of the cranes as they wing their way north over our upturned faces, enacting in rituals the wondrous return of each species from the brink of extinction—while marking in story, and ceremony, the demise of those species we let slip from the real, so that we'll remember, collectively, the awesome cost of our hubris. These are ancient and elemental ways to awaken our own animal senses, binding the imagination of our bodies back into the wider life of the animate earth.

Of course, shortening the long and complex trajectory that food takes from the soil to our mouths—backing off from buying so many fruits imported from elsewhere, while learning to savor the tastes of what grows in our region—this is another, obvious way of bringing our body's imagination back to the living land where we dwell. Growing some food ourselves provides a fuller grounding, and a steady opening of the senses. The swelling number and scale of farmers' markets (and the steady surge of participation in community-supported agriculture cooperatives, or CSAs), indicates that a renewal of place-based intelligence is already under way. Farmers' markets are even now seeding a resurgence of oral culture, for unlike supermarkets, where customers barrel down the aisles enclosed in a solipsistic exchange with their own shopping lists, at farmers' markets most people stroll and linger, bumping into old friends and new, catching up on the gossip even as they compare turnips and exchange recipes. Slowing down, checking out which fruits are finally ripening in the region, tapping our toes to the lilt of a fiddle or an accordion, offering a bag of apples from the backyard in exchange for a braid of garlic and a fistful of cilantro: the convivial ethos of the farmers' market loosens our cynical armor, opening our organism onto the tones and textures of an evolving local vernacular.

We replenish that vernacular by listening deeply and replying in kind. Cultivating oral culture means that we take a new pleasure in simple conversation, becoming gradually more aware of the sonorous qualities of our voice and the audible sound-spell of our speaking. Can we converse with one another less as abstract minds, and more as corporeal, earthly beings engaging other denizens of the earth? A new (and yet in some sense, a very old) attention to the visceral resonance of one's talking, to the rhythm of words and the music of particular phrases, is an obvious way by which to speak in accordance with one's senses, a way to talk without splitting oneself into a speaking intelligence up here and a mute body down there.

Such attunement to the melodic quality of spoken language also draws our interlocutors into alignment with their own senses, and it has this curious correlate: it opens our ears to the expressive intelligence of other sonorous creatures, who generally don't speak in words.

Even as the written word and its conventions came to dominate all our discourse, our anarchic oral eloquence quietly kept itself alive through various means, including the careful and patient craft of literary poets. A renewal of oral culture in the coming era may mean that poetry increasingly frees itself from the printed page and the digital screen, to become a spontaneous part of every person's practice. Poetics, in this sense, would become the practice of alert, animal attention to the broader conversation that surrounds—to the utterances of sunlight and water and the thrumming reply of the bees, or the staccato response of a woodpecker to the hollow creaking of an old trunk—and the attempt not to violate this wider conversation every time that we speak, but to allow it, to acknowledge it, and sometimes to join it.

Renewing oral culture is thus not at all a matter of "turning back the clock," but rather of stepping, now and then, out of clock time entirely. It is not a matter of "going back" to an earlier way of life, but of aligning ourselves with the full depth of the present, expanding awareness beyond the gleaming veneer of our mass-produced artifacts, dropping our attention beneath the recently sedimented strata of commercial civilization (beneath the inert, plastic layers of tossed-out toys and discarded water bottles) to make conscious contact with the darker humus in which our humanity is still rooted. The soil at that depth is made of dances, and songs, and the hushed cadence of spoken stories. By remembering ourselves at that depth, by tapping the nutrients in that timeless soil, we draw fresh water on up into the stems and leaves of the open present. We re-create civilization by tapping the primordial wellsprings of culture, replenishing the practice of wonder that lies at the indigenous heart of all culture.

All our tools of communication will undoubtedly undergo dramatic transformations in the course of the coming decades. Whether printed words such as these will exist at the end of this century is anyone's guess. For the moment, we can be ardent readers (and writers) of books, while recognizing that this abstract and almost exclusively human layer of culture will never be sufficient unto itself. Without rejecting this rich form of interchange, we can nonetheless discern, today, that the rejuvenation of oral culture is an ecological imperative.

CONCLUSION

At the Heart of the Heart of the World

According to a tale once heard in variant forms among numerous placed-based peoples—among Aboriginal elders in Australia and native storytellers in North America, among the old Mayans of Mesoamerica and the ancient Egyptians along the Nile—the fiery sun journeys, every night, through the dark earth underfoot. At dawn we see the sun surface from the ground far to the east, to begin its long, arcing voyage across the vault of sky, lighting the world and feeding the land with its warmth. At evening's approach we watch the sun slide down into the western earth (or slowly immerse itself in the western ocean). Throughout the night, then, the luminous sun walks through the thick darkness that undergirds our lives, firing the material depths with its radiance, pausing in the middle of the night to rest and replenish itself at the center of the earth, before making its way toward the eastern place of its eventual emergence. During the summer, with its drawn-out days, the sun has little chance to tarry in the ground, but in the autumn it hastens more swiftly across the firmament, yielding further time to rest in the rocky density below. Then, at last, during the long nights of winter, and especially at the winter solstice, the sun lingers and sleeps in at the heart of the earth, nourishing the dark ground with its lustrous dreams, infusing the depths with the manifold life that will soon, after several moons of gestation, blossom forth upon earth's surface.

It is a story born of a way of thinking very different from the ways most of us think today. A story that has, we might say, very little to do with "the facts" of the matter. Yet the tale of the sun's journey within the earth holds a curious resonance for many of us who hear it, despite our awareness that the events it describes are not literally true. For the story brings us close to our senses, and to our direct, bodily awareness of the earthly cosmos.

In many versions of the tale, the sun is male; in others, the sun is female. In some tellings, the sun is carried on a boat through the underworld. In others, the sun almost dies during the transit through the gloom, fading to a gaunt skeleton; every dawn it must be clothed anew with flesh and blood in order to make the trek across the heavens. In some stories the setting sun is swallowed into the belly of a huge fish, who then swims with it far to the east, vomiting out the sun at dawn like Jonah from the belly of the whale. Among one tribe of Paiute Indians, the sun who sleeps deep within the earth is a great lizard who eats the stars for nourishment; that is why the stars flee whenever he climbs back into the sky. The sun can't help it: he must swallow some gleaming stars in order to live, and it is they who provide his radiance. The moon, meanwhile, is the wife of that lizard sun, and the mother of those stars. She, too, journeys into the ground to sleep in the house at the heart of the earth. But when her husband arrives home, the moon will sometimes depart—since if the sun has not been able to catch any stars, he will likely be in a lousy mood. The stars are at ease when their mother the moon is in the sky; they sparkle and dance as she passes by. Nonetheless, every twenty-nine days the sun *does* manage to eat a clutch of those stars. Then the moon slowly puts black pitch on her face, in mourning—much as Paiute women darken their faces with pitch when a child dies. After some time the pitch wears off, and the face of the moon becomes visible in its fullness—slightly smudged, perhaps, but as brilliant as we're likely to see her before her husband swallows some further stars and sends her back into mourning.

We can be fairly certain that such oral stories were not taken literally by those who told them, or even by those who heard them. For literal truth is a very recent invention, brought into being by alphabetic literacy. The word "literal," after all, derives from the Latin word for letter. To understand something *literally* originally meant to understand that it happened exactly *as written in scripture.* However, the stories proper to most indigenous cultures—like the diverse creation stories told among tribal peoples throughout the world, or the various tales of the sun's night journey beneath the ground—were commonly told and retold without ever being written down.

Once inscribed on the page, language—as we've learned—takes on a detached fixity very different from the way language is experienced in a deeply oral culture. To our unaided, animal senses, everything speaks. But with the spread of phonetic writing, the capacity for meaningful speech seemed to withdraw from the surrounding landscape; language increasingly came to seem a purely human prerogative. Words and phrases began to be used less as invocational and creative powers, and more as representational functions—as labels by which to demarcate and define a mute cosmos that had previously appeared animate, and expressive, in its own right. Material things seemed to become more stable and determinate.

Only after the advent of written letters, and the dissemination of alphabetic texts, did there arise a clear distinction between "literal" and "metaphoric" uses of language (between literal truth and figurative, or poetic, truth). For this simple reason, the many indigenous tales of the sun's night journey through the ground could hardly have been understood *literally* by those who heard them in their original ambiance.

We would be similarly mistaken if we concluded that such stories carried a purely *metaphoric* meaning for those who originally told and heard them. For again, these tales were being told long before that writing-induced rift in felt meaning that forged the notion of *literal speech,* on the one hand, and *metaphoric speech,* on the other. We must try to imagine ourselves into a mode of listening prior to any such split if we wish to hear the old, ancestral stories in anything like their original meaning.

When we say that a statement is "purely metaphoric," we often mean that it is an imaginative formulation that has little direct bearing upon the tangible world. Yet the old stories were the primary linguistic form by which to convey practical information regarding the tangible cosmos; it's clear that oral peoples understood these stories to be about the actually experienced world. If the stories referred to *the underworld,* this could not have been a "merely metaphorical" realm. If the underworld was hidden from us humans, it was nonetheless a hidden aspect of *this very palpable terrain.* It was concealed from our awareness by the opacity of the solid ground beneath our feet—by this very ground crisscrossed by the tracks of moose and wolf and bear, into which we dug for specific roots and upon which we laid down to sleep. The underworld, in other words, was precisely under the ground.

And yet, again, this selfsame sensuous terrain could not have been experienced as a *literal* reality: it could not have been a world of determinate and stable facts. Neither entirely literal nor entirely metaphoric, the world that articulated itself through our oral stories was rather, at every point, *metamorphic.* The land was alive: each place had its pulse, each palpable presence seemed crouched in readiness to become something else.

Such is the realm opened to us, still, by our animal senses. Our most immediate perceptual experience discloses a world in continual metamorphosis. Even the most allegedly stable landforms alter around us as we move among them, their hues transforming as the sun glides behind the clouds. The tonalities of each region modulate with the turning seasons. Two weeks ago, when my partner and I were gathering wild herbs in the mountains north of our home—digging osha roots and plucking nettles—I heard the very faint but unmistakable call, somewhat like the rusty hinge of an old screen door, of a sandhill crane. I glanced up into the cloudless, clarion blue of the autumn sky, but could see no bird. I scanned the surrounding valley and the cliffs looming above us. Nothing. Then again I heard it—that rusty but evocative bugling, seeming to come from several directions at once, as if there were several cranes, and it occurred to me that it was echoing off the tall cliffs. I stared back up into the sky, and suddenly far, far up there a glimmering white pattern crystallized out of the blue. It was a perfect V-shaped arrowhead, made of thirty or thirty-four cranes, visible only as the sun reflected off their flapping wings, a ripple of white spreading from the point backward along each slanted edge as the arrow advanced across the heavens. I stared and stared until the apparition was directly overhead, then took a moment to glance at Carmen; she looked back at me, grinning, and we swiveled our faces back toward the sky. Except . . . where were they? I poured my gaze into every part of the sky's expanse, but could not find that slow-motion arrow. The cranes had dissolved back into the blue depths.

I heard Carmen's voice, somewhat anxious: "Can you see them?"

"No."

"Did they just disappear, or what?!"

Two or three times, a bugling cry stuttered down out of the heights, but our eyes were unable to overcome the sky's witchery, and we finally gave up.

The morning after that apparition and odd vanishing, I was searching for more roots when I was spooked by a brazen animal who chased me, leaping and crashing, down the wooded slope—my adrenaline gushing—until the predator abruptly resolved into a large, dislodged rock that tumbled on past. Relieved and shaken, I stopped to catch my breath. As my pulse eased back down, I noticed an abundance of dazzling cerulean blossoms on a nearby bush. Unable to identify them, I stepped closer to better inhale their color with my eyes, yet at my approach the blossoms seemed to quiver and undulate, then all at once they flapped skyward, morphing into a flock of blue butterflies.

Reality shapeshifts. Underneath our definitions, prior to all our ready explanations, the world disclosed by our bodily senses is a breathing cosmos—tranced, animate, and trickster-struck.

The oral stories, then, bring us close to our animal senses. They recall us to our bodily participation in the metamorphic depths of the sensuous. The tale of the sun's journey through the ground stirs and resonates within us because it cuts through our easy abstractions, and calls us back to our most direct, creaturely encounter with the space around us. Our spontaneous, sensory awareness of the sun is of a fiery presence that rises and sets. Despite all we may have learned about the stability of the sun relative to the earth, no matter how thoroughly we've convinced our intellects that it is the earth that is really turning while the sun basically holds its place, our animal eyes still perceive the sun mounting up from the distant earth every morning, and sinking beneath the far-off ground every evening. Whether we are farmers or physicists, we all speak of the "rising" and the "setting" of the sun, for this remains our primary experience of the matter.

The story follows a kind of perceptual logic very different from the abstract logic we learned at school. It attends closely to the sensuous play of the world, allowing the unfolding pattern of that display to carry us into a place of dark wonder and possibility: that at night the sun replenishes itself in the material depths of the ground. There is a vivid imagination at work in the tale, although it's an imagination steadily nourished by our senses, and one that nourishes them in turn. The story does not ask us to forsake the evidence of our eyes, but invites us to look deeper, and to listen ever closer, feeling our way into participation with a palpable cosmos at least as alive and aware as we are. The jostling elemental powers that compose this animate cosmos are sometimes lucid and sometimes dazed—like us, they must give themselves over to sleep, and the magic of dreams, if they wish to renew themselves.

Informed by the logic of our creaturely senses, the story gestures toward a great secret: that there's a blazing luminosity that resides at the heart of the earth. The tale suggests that the salutary goodness of light makes its primary home within the density and darkness of matter. That the transcendent, life-giving radiance that daily reaches down to us from the celestial heights also reaches up to us from far below the ground. That there's a Holiness that dwells and dreams at the very center of the earth.

That which transcends the sensuous world also secretly makes its home deep *within* this world. However blasphemous such an affirmation may sound to persons of a theistic bent, the aboriginal intuition of a resplendence immanent in matter accords well with a new sense of the sacred now striving to be born.

Our age-old disparagement of corporeal reality has in our time brought not just our kind but the whole biosphere to a horrific impasse. The aspiration for a bodiless purity that led so many to demean earthly nature as a fallen, sinful realm (and the related will-to-control that's led us to ceaselessly mine and manipulate nature for our own, exclusively human, purposes) has made a mangled wreckage of this elegantly interlaced world. Yet a new vision of our planet has been gathering, quietly, even as the old, armored ways of seeing stumble and joust for ascendancy, their metallic joints creaking and crumbling with rust. Beneath the clamor of ideologies and the clashing of civilizations, a fresh perception is slowly shaping itself—a clarified encounter between the human animal and its elemental habitat.

It is a perception that honors the immeasurable otherness of things, the way that any earthborn presence exceeds the calculations we perform upon it—the manner in which each stone, each gust of wind, each termite-ridden log or gliding sea turtle harbors and bodies forth a creativity that resists all definition. As though there's a subtle fire burning within each sensible presence, a heartbeat unique to each being—not only to persons, then, and individual woodpeckers, but also bulrushes and granite slabs, gashes burned into trees by lightning, pollen grains, katydids, coral reefs, and shed snakeskins. This unique creativity ensures that we don't really perceive the beings around us unless we suspend our already-settled certainties, opening ourselves toward whatever pulse rides within each thing we meet. The expectation of a basic enigma at the heart of every ostensible "object" kindles a new humility within ourselves, engendering an empathic attunement to our surroundings and a compassionate resolve to do least harm.

Despite our inherited conceptions, sensible things are not fixed and finished objects able to be fathomed all at once. Their incomplete quality opens them to the influence of other things, ensuring that each entity—earthworm, musk ox, thundercloud, cactus flower—is held within an interdependent lattice of relationships, a matrix of exchanges and reciprocities that is not settled within itself but remains fluid and adaptable, able to respond to perturbations from afar, yielding a biosphere that is not, finally, a clutch of determinate mechanisms but a living sphere, breathing . . .

If much natural science of the last two centuries held itself aloof from the nature it studied, pondering the material world as though that world were a huge aggregate of inert objects and mechanical events, many new-age spiritualities simply *abandon* material nature entirely, inviting their adherents to focus their intuitions upon non-material energies and disincarnate beings assumed to operate in an a-physical dimension, pulling the strings of our apparent reality and arranging earthly events according to an order that lies elsewhere, behind the scenes. Commonly reckoned to be at odds with one another, conventional over-reductive science and most new-age spiritualities actually fortify one another in their detachment from the earth, one of them reducing sensible nature to an object with scant room for sentience and creativity, the other projecting all creativity into a supernatural dimension beyond all bodily ken.

A similar alliance, unsuspected by those most caught within it, may be found in the contemporary ideological battle between the advocates of creationism (or, as many currently frame themselves, the proponents of "intelligent design") and the neo-Darwinian dogmatists of the "new atheism." The intelligent design theorists insist that many aspects of present-day organisms are too complex, and too perfectly adapted for the particular functions that they serve, to have arisen only as a result of the undirected or "blind" evolutionary process of natural selection acting upon purely random mutations. Such instances of irreducible complexity, they maintain, can only have been designed by an external intelligent designer—by a God exerting His will upon the material cosmos from outside.

It is true that countless attributes of organisms are exquisitely tuned for the roles they currently play (the eye of the raptor, the lungs of a sperm whale, the insect-mimicking parts of certain orchids, the human capacity to compose love poems), so much so that a purely haphazard process of mutation, blindly sorted over the millennia by an indifferent environment, seems almost absurd as a description of how such arrangements came into being. Nonetheless, many natural scientists, more doctrinaire than Darwin himself, insist that evolution must be interpreted in this curiously cramped manner, as a thoroughly oblivious process whereby, in the competition for limited resources, a strictly indifferent and insentient nature exerts selective pressure upon purely random genetic mutations. Never mind that many of the beings that compose the environment exhibit obvious signs of sensitivity and sentience: whenever we reflect upon the evolutionary process we are constrained to consider the apparent agent of selection—the earthly environment, or biosphere—as a purely objective, quantifiable set of conditions without any subjectivity or creativity. In order to do so, however, we must subtract our own sentience (and that of many other beings) from the earthly surroundings. Since it is our own sentience or subjectivity that is engaged in pondering the evolutionary process, we must somehow locate our pondering selves elsewhere, must situate our sentience outside the biosphere, looking upon earthly nature from a disinterested position radically apart from that nature. In this we are aided, immeasurably, by our civilization's age-old faith in an all-seeing God, a divinity that ponders the material world from a position wholly external to that world. The scientific intellect, which sometimes prides itself on having vanquished the belief in God from much of the rational populace, regularly situates its gaze in the very place (or rather, the very same *non-place*) recently vacated by that God. For it affects the same external, all-seeing perspective, the same *view from nowhere* enjoyed by that divinity. The most assertive new atheists unwittingly rely, in this sense, upon the very same monotheistic assumptions that they ostensibly oppose.

The hyper-rational objectivity behind a great deal of contemporary techno-science could only have arisen in a civilization steeped in a dogmatic and other-worldly monotheism, for it is largely a continuation of the very same detached and derogatory relation to sensuous nature. If in an earlier era we spoke of the earthly world as fallen, sinful, and demonic, we now speak of it as mostly inert, mechanical, and determinate. In both instances nature is stripped of its generosity and prodigious creativity. Similarly, the utopian technological dreaming that would have us bioengineer our way into a new and "more perfected" nature (or would have us download human consciousness into "better hardware"), like the new-age wish to spiritually transcend the "physical plane" entirely, seems calculated to help us hide from the shadowed wonder and wildness of earthly existence.

All of these dodges, all of these ways of disparaging material nature or of aiming ourselves elsewhere, enable us to avoid the vulnerability of real relationship with other persons and places in the depths of this unmasterable world. Despite the several pleasures we might draw from life in this world, there remains something about earthly reality that frightens us, and especially unnerves most of us born into civilization. Not just the decay to which our earthborn bodies are prone, and the death that patiently awaits us, but also our steady subjection to what exceeds us, to the otherness of other persons and other beings, and to an anarchic array of elemental forces over which we have little control. To exist as a body is to be constrained from being everything, and so to be exposed and susceptible to all that is not oneself—able to be tripped up at any moment by the inscrutability of a pattern one cannot fathom.

Whether sustained by a desire for spiritual transcendence or by the contrary wish for technological control and mastery, most of our contemporary convictions carefully shirk and shy away from the way the biosphere is directly experienced from our creaturely position in the thick of its unfolding. They deflect our attention away from a mystery that gleams and glints in the depths of the sensuous world itself, shining forth from within each presence that we see or hear or touch. They divert us from a felt sense that this wild-flowering earth is the primary source of itself, the very well-spring of its own ongoing regenesis. From a recognition that *nature,* as the word itself suggests, is self-born. And hence that matter is not just created but also *creative,* not a passive blend of chance happenings and mechanically determined events, but an unfolding creativity ever coming into being, ever bringing itself forth . . .

Why is this simple and rather obvious intuition—this recognition of matter as generative and animate—so disturbing to civilized thought? It's as though there's an ancient dread of what is palpably dense, an old and unspoken taboo against acknowledging the creativity of matter—as if by such a recognition we risk waking a slumbering power that intends us harm. An ancestral sense that whatever is genuinely good in this world must have its ultimate source in what is above and ethereal, while whatever is dense, dark, and downward must be avoided at all costs. As though the damp soil underfoot was solely a medium of death and decay and not, as well, the very source and fundament of new life. As though what is deeper down below is best not pondered at all, lest we fall under its infernal influence. For (and let us hope that we don't provoke its wrath by such speaking) is not that deep-down place the terrible locus of Hell, the very dwelling of Satan and the fiery source of all that's evil?

Is this, then, why we feel compelled to distinguish our reflective selves from our material bodies, and strive to hold ourselves aloof from the density of earth? Is not this civilization, in both its religious and its secular variants, still beholden to an old topological aversion layered deep within our languages and tightly held within our muscles? An inherited, habitual dichotomy between an absolute Good that summons us from on high and a perfect Evil that drags us down?

Whenever the wild diversity of experience is twisted into a simple opposition between what's good and what's bad, whenever the heterogeneous multiplicity of life is polarized into a battle between a pure Good and a pure Evil, then the earth itself is bound to suffer at our hands. When the sacred is conceptually stripped of its various shadows and idealized as a pure light, or Goodness, without any taint of the dark, then those stripped-away shadows inevitably seem to gather into a concentrated and implacable gloom, or Badness. The unsullied light can only be located above and beyond this ambiguous world with its shadowed woodlands and its swamps, its cycles of growth and decay. The unadulterated darkness, meanwhile, must be located in a realm utterly inaccessible to that light, and is therefore assumed to dwell far underground, at the center of the earth. And earthly nature, for all its abundance, comes to seem a tainted place, all too much under the abysmal influence of what lies below. Sure, this world is illuminated by the sun's radiance, by that light that draws new shoots from the soil and beckons our spirits to ascend. But the weight and density of our material bodies render us vulnerable to the pull of what lies below. Our thick physicality holds us to the ground, drawing us down toward that place of sheer dread, without light, at the dense center of this world. There's no escaping this downward drag while we're alive. No wonder a civilization steeped in the polarization of Good versus Evil wreaks such havoc on the rest of nature. No wonder so many creatures are dwindling and disappearing, their homes ravaged with toxins, their forests transformed to stumps . . .

Was this inevitable? The old oral stories about the sun's night journey through the ground give evidence that the simple opposition between an infinite light and a concentrated dark—between a perfect goodness and a pure malevolence—is not instinctive for the human animal, and indeed is of very recent vintage relative to the enormous depth of our indigenous ancestry. The stories suggest, on the contrary, that there is a secret identity between the resplendent sun overhead—whose face we dare not gaze upon directly—and the wild power that resides at the center of the earth. The tales draw their evocative strength, as we've noticed, from the way the solar fire has always been experienced by our unaided, animal senses. Our bodies observe the sun's trajectory as it arcs through the heavens and slides down into the western earth at sunset; they observe the sun climbing out from the eastern lands—with a commotion of color, and much fanfare of birdsong—every dawn. How simple, then, for our sensory imagination to conclude that the sun is traveling through the ground at night, and taking its rest at the center of the earth! How plain and obvious the primordial intuition that the sun dwells, and sleeps, at the heart of the earth!

The intuition, however, appears antithetical to our contemporary understanding of things. It seems especially absurd in light of the Copernican basis of modern science. But it is not absurd. Although couched in a storied way of speaking very different from the cool detachment often promoted by our sciences, the inward sun holds a meaning uniquely resonant with today's insights. Consider: according to our current understanding, the mass of the sun—now recognized as our local star, an outrageously immense ball of flaring gas roughly a million times the size of the earth—holds our planet in distant orbit around itself through the power of gravitational attraction. Yet while the sun exerts its gravitational pull upon the entirety of our planet, that force is not felt equally by every part of the earth. For the mass of the sun focuses its grip upon earth's *center of gravity*. And that center is located deep within the white-hot core of the earth, far beneath the ground on which we stand. The sun's mass, in other words, holds the earth most tightly at earth's center; the rest of our planet spins, like a slowly wobbling top, around that stable center.

Hence, while the sun's extravagant radiance reaches us daily from above, filtering down through earth's atmosphere, the sun's most incessant influence upon our world and upon our lives reaches us from below. It is an influence so visceral and direct that we can hardly bring it into our awareness. It is a connection as constant by day as it is by night, a relationship that does not vary as the cool of morning gives way to the warmth of noontime, or as the glare of midday is muted by the lengthening shadows of dusk. After Newton's discoveries regarding the universal *mutual attraction* (or eros) we now call gravity, we know that our most direct material engagement with the sun is through the gravitational center of our planet. For our muscled, animal bodies, that blazing star makes itself felt not only as it moves across the daytime sky, but more deeply and constantly through the very heart of the earth.

To recount a tale suggesting that the sun that soars exalted through our skies also dwells deep within the earth—that the same round fire that lights our days secretly makes its home far beneath the ground on which we sleep—is hardly, then, to promote an ancient fiction. For the story is a simple and elegant way of translating a founding insight of modern science into the corporeal immediacy of felt experience. By carefully honing and telling such tales as this, we begin to *complete* the Copernican revolution—bringing its insights, at long last, back down to earth.

If the discoveries of Copernicus, Kepler, and Galileo loosened the earth from the center of the universe, they also opened a profound rift between our sensing bodies and our thinking minds—between our direct, sensory experience of the world, in which the sun moved across the sky of a stable earth, and our new intellectual apprehension of the world, in which the sun remained stable while the earth itself moved. If the sun was really motionless, then everything our senses showed us was untrustworthy. As Copernican and Newtonian insights took hold, sensory perception was increasingly derided as deceptive; only that which could be measured and analyzed mathematically could be taken as true. The spreading cultural detachment from bodily experience enabled a new audacity in our human researches, empowering a wondrous range of discoveries and technological innovations. But it also left us curiously adrift, bereft of our most immediate source of contact and rapport with the surrounding terrain. Dismissing our felt experience, we sacrificed much of our animal empathy with the animate earth, forfeiting the implicit sustenance we'd always drawn from that empathy. While amassing our analytic truths and deploying our technologies, we became more and more impervious to the needs of the living land, oddly inured to the suffering of other animals and to the fate of the more-than-human world.

The preceding chapters have tried to delineate some of the dimensions of our perceptual oblivion, exploring an array of ways to recover our attunement without abandoning intellectual rigor. Corporeal sensations, feelings, our animal propensity to blend with our surroundings and be altered by them, our bedazzlement by birdsong and our susceptibility to the moon: none of these ought to be viewed as antithetical to clear thought. Our animal senses are neither deceptive nor untrustworthy; they are our access to the cosmos. Bodily perception provides our most intimate entry into a primary order of reality that can be disparaged or dismissed only at our peril. Far from offering an untrustworthy account of things, our senses disclose an ever-shifting reality that is not amenable to any finished account, an enigmatic and encompassing field of relationships to which we can only apprentice ourselves. This ambiguous order cannot be *superseded* by reason and the careful practice of our sciences, since it provides the experiential substance without which reason becomes rudderless. As the very substance of the real, it cannot be supplanted—but it can be augmented, elaborated, clarified, and complexified by those sciences. And our participation *within* it can be honed and deepened by our discoveries.

One such discovery was the recognition by the astronomer William Herschel some two hundred years ago that, contrary to the assertions of Copernicus, the sun itself was not stable but in motion, drawing the Earth and other planets with it as it whirls around the center of the Milky Way galaxy. Subsequent discoveries have confirmed that that same galactic center—like the center of every other galaxy—is itself in motion through space. However, when we say that something is in motion, we generally mean that it is moving relative to some stable point, some unmoving ground or medium. In a cosmos where every presence is motion, what shall we choose as that still point? Shall we designate our own sun as the stable reference? Or perhaps the center of some other solar system? Shall we choose the center of our galaxy, or the gravitational center of the great cluster of galaxies in which our Milky Way resides? But that center, too, is in motion relative to other clusters. In truth, *any* cluster or galaxy or star may be chosen as the stable reference, in relation to which every other body is seen to be in slow or rapid motion. In such an unbounded and dizzying pluriverse as ours, teeming with uncountable galaxies, *every* sphere enacts a center around which all the rest arrays itself. Why not, then, our own sphere, our own wild-flowering Earth?

Everything that we animals understand of movement and stillness was first acquired from our bodily experience of being in motion or at rest relative to the ground of this Earth, learned from crawling and walking upon it, and curling to sleep on its ample expanse. If we are told, for example, that some location at the center of the Milky Way is basically at rest while the remainder of our galaxy whirls around it, we conceptualize that center only by transferring our felt, bodily experience of the stable ground over into that presumed place of stillness. Without such an associative transfer, the so-called stability of that center could have no meaning for us. We cannot feel the relative rest or repose of any other cosmic location without first assimilating Earth's ground (our body's inescapable frame of reference) to that other location.

Even if we choose to say, with Copernicus, that it is not the Earth but rather the sun that is really at rest, we cannot help but approach that stable sun through the abiding, earthly ground beneath our feet.

Tonight is the winter solstice, the dark of the year. This book is completed. Too many species have slid into extinction during the writing, too many forests felled and wetlands filled; so much beauty's fled the world. Life's become cheap: with more and more of us piling in, humans keep bashing each other in ever more creative ways—car bombs bursting bodies and missiles dropped from unmanned drones splintering families, searing the land and spattering it with blood. An addled and anesthetized numbness is spreading rapidly throughout our species.

There are those, however, who are not frightened of grief; dropping deep into the sorrow, they find therein a necessary elixir to the numbness. When they encounter one another, when they press their foreheads against the bark of a centuries-old tree, or their palms into the hand of yet another child who has tasted prematurely of wrenching loss, their eyes well with tears that fall easily to the ground. The soil needs this water. Grief is but a gate, and our tears a kind of key opening a place of wonder that's been locked away. Suddenly we notice the sustaining resonance between the drumming heart within our chest and the pulse rising from under the ground.

The stars glimmer in the solstice dark, their faint light mirrored in glints off the crusted snow. Far below these blanketed fields, deep beneath the bedrock, a lustrous power slumbers, fitfully, like a bear in its cave. The resplendence it carries by day is now subdued and smoldering—a slow burn, crackling within its hearth at the heart of the Eairth. As this power sleeps, it dreams. The dreams roil and flicker and seethe, curling back upon themselves and sometimes flaring, scorching the walls and scattering sparks. A few sparks embed themselves like seeds in the enfolding dark, others wink out and vanish. Meanwhile, the power sleeps, pulsing like a muscle, its vigor radiating outward in waves through the viscosity of molten metal and the slow solidity of rock (firing huge convection currents within these depths that move the continents far above), percolating outward as magma or propagating upward through the density of basalt and granite, rising later through thickets of feldspar and quartzite and the stratified soils near the surface, channeled outward through stems of dandelions and trunks of sequoias, through cattails and sugar maples and the upright backbones of smooth-skinned primates, finally fountaining into the open biosphere through blossoms and budded leaves and through the craft of our fingers, through the gleam in your lover's eye and the fluted music upwelling now from the beak of a blackbird . . .

ACKNOWLEDGMENTS

A deep bow of gratitude is due to Carmen Harris, my gentle ally and intrepid partner in all manner of unhinged adventures. Carmen's curious and creative spirit enlivened many of the questions that lope through this book, and her attentive reading of my words empowered me to hear them in new ways. She is aspen, seashell, and lightning rolled into one being, lithe and beloved.

My two children, Hannah Cecilia Night and Leander Night Rivers, have been especially patient with me as I wrestled with this work. Their capacity for joy and their pleasure in sharing it, their rapt fascination at hearing a tale well told, their exhilaration at accompanying their dad on yet another tramp though the woods—for all this and much more, for simply being who they are, I thank them from the depths of my heart.

I'm grateful, as well, to the soulful intelligence of the three mischievous colleagues with whom I founded the Alliance for Wild Ethics: Stephan Harding, Per Ingvar Haukeland, and Per Espen Stoknes. This book has gained immensely from our chances to teach together, and from our several gatherings in the backcountry—from the opportunity to pool our intuitions and test our insights in interaction with the thunderstruck mountains.

Grietje Laga was my companion during the initial phase of this project; she remains a close comrade and friend.

For careful reading of various chapters, and keen feedback on the ideas therein, I am greatly indebted to systems theorist Stuart Cowan, as well as to Geneen Haugen, Ann Hunkins, Stephanie Mills, and my bodacious and brilliant friend Ben Gillock. Mythologist Sean Kane (from whom I stole the phrase "the real in its wonder") was a marvelously attentive and ruthless reader. A couple warm dialogues with physicist Brian Swimme helped hone certain reflections, as did conversations with a broad range of luminous souls, among them Jay Griffiths, Kalevi Kull, Wendell Berry, David Cayley, Patrick Curry, Donna House, Omar Zubaedi, Greg Glazner, Eva Simms, Tom Jay, Will Adams, Niel Thiese, Morten Tonnessen, Stefan Laeng-Gilliatt, Maya Ward, Jan van Boekel, Deborah Bird Rose, Ed Casey, Bill Plotkin, Keren Abrams, Steve Talbott, Craig Holdrege, Eileen Crist, the late and much-missed Brian Goodwin, Peter Adams, Chris Wells, Jennifer Sahn, Jon Young, Peter Manchester, and Arthur Zajonc.

The Levinson Foundation and the remarkable Lannan Foundation provided generous support at crucial moments; they have my profound gratitude.

Dan Frank is a sharp editor and a delightful guy, gracefully coaxing and cajoling many fine writers into print. I am also mighty thankful for the care and patience of Jillian Verrillo at Pantheon, and for the diligent craft of my agent, Joe Spieler.

I extend my appreciation to herbalists Peggy Creelman and Suzanne Barry, for skillful support during my long grapple with Lyme disease.

Some of the explorations in this book drew their initial impetus from investigations undertaken in the mid-twentieth century by the French philosopher and phenomenologist Maurice Merleau-Ponty; I remain indebted to his ethical and engaged thinking.

While writing these pages, two wise old friends slipped back into the wider life of the planet. Theologian turned geologian Thomas Berry had long fed my life with his radical wonder for the terrestrial ground of our experience. Mountain philosopher Arne

Naess touched all around him with his playful and prodding clarity, and his steady allegiance to what mattered. Although very different from one another, each of these elders was uniquely open to the shadowed ambiguity of earthly existence, and each strove to embody this richness in a way of living and of thinking under the elemental influence of that which exceeds our species. Like many other elders still among us, these two creatures were a boon for our biosphere, scattering the seeds of a wild culture more awake to our embedment within a more-than-human cosmos.

Finally, the kindness and artistry of my parents, Blanche Abram and Irving Abram—the steady warmth of their friendship, and the convivial community of spirited folks that gather around them—has remained a touchstone for my life; my gratitude to them will not fade. There are so many unsung heroines and heroes at this broken moment in our collective story, so many courageous persons who, unbeknownst to themselves, are holding together the world by their resolute love or their contagious joy. Although I do not know your names, I can feel you out there.

A NOTE ON SOURCES

The poem by Robert Bringhurst is "A Quadratic Equation" and can be found in Robert Bringhurst, *Selected Poems* (Kentville, Nova Scotia: Gaspereau Press, 2007).

The John Muir quote on page 81 is from a journal entry Muir wrote in 1913, the year before he died. See Linne Marsh Wolfe, ed., *John of the Mountains: The Unpublished Journals of John Muir* (Boston: Houghton Mifflin, 1938), p. 427.

Gertrude Stein's words on page 131 are from her 1936 lecture "An American and France." See Gertrude Stein, *The Making of Americans: Being a History of a Family's Progress* (Champaign, Ill.: Dalkey Archive Press, 1995), p. xvi.

The words of the Carrier Indian on page 259 are quoted in Diamond Jenness, *The Carrier Indians of the Bulkley River,* Bureau of American Ethnology Bulletin 133 (Washington, D.C.: Smithsonian Institution, 1943), p. 540.